4th Workshop on Theory, Modelling and Computational Methods for Semiconductors

(TMCSIV 2014)

Journal of Physics: Conference Series Volume 526

Salford, United Kingdom
22 – 24 January 2014

Editors:

J. Pal
M. Migliorato
S. Tomic

ISBN: 978-1-63439-307-2
ISSN: 1742-6588

Printed from e-media with permission by:

Curran Associates, Inc.
57 Morehouse Lane
Red Hook, NY 12571

Some format issues inherent in the e-media version may also appear in this print version.

Printed by Curran Associates, Inc. (2014)

Institute of Physics
Dirac House, Temple Back
Bristol BS1 6BE UK

Phone: 44 1 17 929 7481
Fax: 44 1 17 920 0979

techtracking@iop.org

Additional copies of this publication are available from:

Curran Associates, Inc.
57 Morehouse Lane
Red Hook, NY 12571 USA
Phone: 845-758-0400
Fax: 845-758-2634
Email: curran@proceedings.com
Web: www.proceedings.com

TABLE OF CONTENTS

Journal of Physics: Conference Series **526** (2014) 011001 doi:10.1088/1742-6596/526/1/011001

Preface

These conference proceedings contain the written papers of the contributions presented at the 4th International Conference on Theory, Modelling and Computational Methods for Semiconductor materials and nanostructures. The conference was held at the MediaCityUK, University of Salford, Manchester, UK on 22–24 January 2014. The previous conferences in this series took place in 2012 at the University of Leeds, in 2010 at St William's College, York and in 2008 at the University of Manchester, UK.

The development of high-performance computer architectures is finally allowing the routine use of accurate methods for calculating the structural, thermodynamic, vibrational, optical and electronic properties of semiconductors and their hetero- and nano-structures. The scope of this conference embraces modelling, theory and the use of sophisticated computational tools in semiconductor science and technology, where there is substantial potential for time-saving in R&D. Theoretical approaches represented in this meeting included: Density Functional Theory, Semi-empirical Electronic Structure Methods, Multi-scale Approaches, Modelling of PV devices, Electron Transport, and Graphene. Topics included, but were not limited to: Optical Properties of Quantum Nanostructures including Colloids and Nanotubes, Plasmonics, Magnetic Semiconductors, Photonic Structures, and Electronic Devices. This workshop ran for three days, with the objective of bringing together UK and international leading experts in the theoretical modelling of Group IV, III-V and II-VI semiconductors, as well as students, postdocs and early-career researchers. The first day focused on providing an introduction and overview of this vast field, aimed particularly at students, with several lectures given by recognized experts in various theoretical approaches. The following two days showcased some of the best theoretical research carried out in the UK in this field, with several contributions also from representatives of renowned theoretical groups from many European countries (Spain, France, Ireland, Germany, Switzerland, Luxemburg, Norway, Italy, Poland, Denmark, Sweden, Serbia, etc.), as well as Asia (Iran, Japan) and USA.

We would like to thank all participants for making this a very successful meeting and for their contribution to the conference programme and these proceedings. We would also like to acknowledge the financial support from the Institute of Physics (Semiconductor Physics Group and Computational Physics Group), EPSRC-UK, the CECAM UK-Hartree Node, CCP9, and Quantum Wise (distributors of Atomistix).

The Editors

Journal of Physics: Conference Series **526** (2014) 011001 doi:10.1088/1742-6596/526/1/011001

Acknowledgments

Conference Organising Committee:

Stanko Tomić (Chair, University of Salford)
Matt Probert (University of York)
Max Migliorato (University of Manchester)
Joydeep Pal (University of Manchester)

Programme Committee:

- David Whittaker (University of Sheffield, UK)
- John Robertson (University of Cambridge, UK)
- Risto Nieminen (Helsinki University of Technology Finland)
- Eoin O'Reilly (Tyndall Institute Cork Republic of Ireland)
- Marco Califano (University of Leeds, UK)
- Stewart Clark (University of Durham, UK)
- Stanko Tomić (University of Salford, UK)
- Mauro Pereira (Sheffield Hallam University, UK)
- Aldo Di Carlo (University of Rome "Tor Vergata," Italy)
- Lev Kantorovich (King's College London, UK)
- Mervin Roy (University of Leicester, UK)
- Ben Hourahine (University of Strathclyde, UK)
- Rita Magri (University of Modena and Reggio Emilia, Italy)
- Zoran Ikonic (University of Leeds)
- John Barker (University of Glasgow)

The proceedings were edited and compiled by Joydeep Pal, Max Migliorato and Stanko Tomić.

Peer review statement

All papers published in this volume of *Journal of Physics: Conference Series* have been peer reviewed through processes administered by the proceedings Editors. Reviews were conducted by expert referees to the professional and scientific standards expected of a proceedings journal published by IOP Publishing.

Journal of Physics: Conference Series **526** (2014) 012001 doi:10.1088/1742-6596/526/1/012001

Causal self-energies for NEGF modelling of quantum nanowires

J Barker[1], A Martinez[2], M Aldegunde[2] and R Valin[2]

[1]School of Engineering, University of Glasgow, G12 8LT, UK,
[2]College of Engineering, University of Swansea, UK.

Email: John.Barker@glasgow.ac.uk

Abstract. Electron-phonon scattering is shown to increase dramatically at small nanowire cross-sections so that the transport is not ballistic. Non-ballistic dissipative device modelling requires the full complexity of the non-equilibrium Green Function (NEGF) method. The role of causality in obtaining spectral sum rule-conserving approximations to the electron-phonon self-energies is demonstrated and applications given for wrap-round gate silicon nanowire field effect transistors. Causality violations give erroneous density of states and current densities.

1. Introduction

Non-Equilibrium Green Function (NEGF) modelling is important for studying future limiting scale transistors such as the gate-all-around nanowire field effect transistor (GAA NWFET [1]). Such devices (Fig.1) were once thought to exhibit *ballistic* transport [2-3] that requires only a simple NEGF formalism. However, from our NEGF simulation studies of phonon-limited mobility shown in Fig. 2, including other data, it is clear that for *small* NW cross-sections (1-3 nm) the inelastic electron-phonon interactions increase significantly (due primarily to the dependence on the quasi-1D density of states). The transport becomes *dissipative* with a significant degradation in effective mobility [3-5].

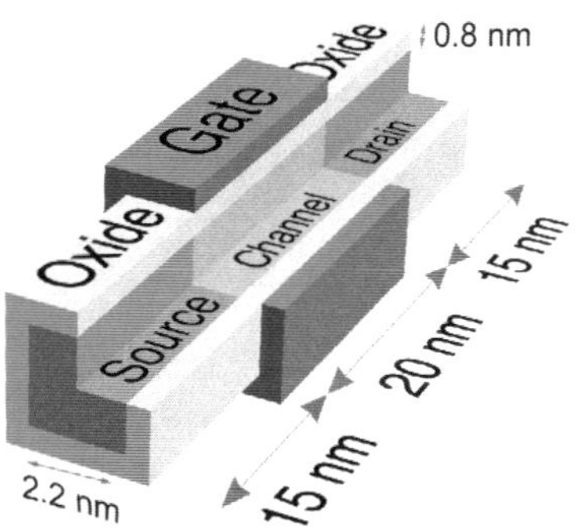

Figure 1: Schematic of a Si Gate-All-Around Nanowire Field Effect Transistor.

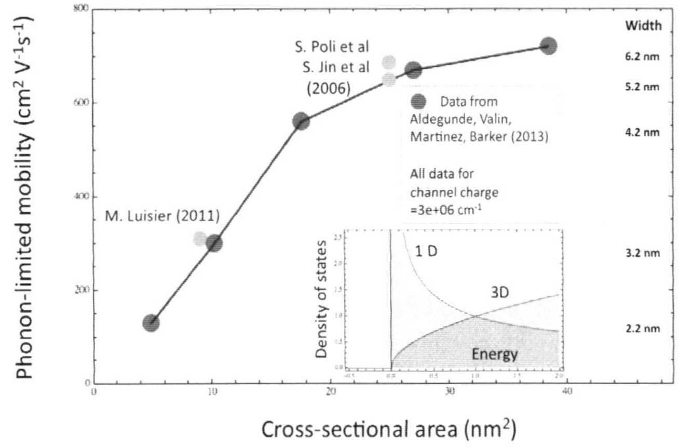

Figure 2: Phonon-limited mobility in Silicon Nanowires as a function of cross-sectional area of the silicon channel.

Computing the self-consistent self-energies for inelastic scattering is problematic, particularly for non-local self-energies. Traditionally, the computation may be lessened by first computing $\Gamma(E)$, the imaginary part of the retarded self-energy $\Sigma(E) = \Delta(E) - i\Gamma(E)$ and using a Kramers-Kronig dispersion relation to compute the real part $\Delta(E)$. Often $\Delta(E)$ is neglected: we show here that this strategy may lead to serious errors in the local density of states (DOS) and estimates of dissipation.

Journal of Physics: Conference Series **526** (2014) 012001 doi:10.1088/1742-6596/526/1/012001

2. Theory

The NEGF method computes the $G^<(r_1, r_2; E)$ Green function which determines the carrier density and the current density as functions of position. For a stationary system $G^<$ is determined from two coupled equations: Dyson's equation for the retarded Green function G^R

$$\{E - i\eta - H_0 - \Sigma^R\}^{-1} G^R = 1 \tag{1}$$

(where the causal factor $\eta \to 0^+$) and the quantum kinetic equation for $G^<$:

$$G^< = G^R \Sigma^< G^A \tag{2}$$

The total self-energy Σ^κ (κ = <, R, $A = advanced$) comprises terms due to the contact regions (S=*source* and D=*drain*) and a term due to the electron-phonon interactions (I):

$$\Sigma^\kappa = \Sigma_S^\kappa + \Sigma_D^\kappa + \Sigma_I^\kappa \tag{3}$$

The Ward identities that enforce the local conservation laws (including current conservation) require at least that Σ_I^κ satisfies the self-consistent Born approximation (SCBA). For inelastic inter-valley optical phonon scattering with phonon energy $\hbar\omega$ the SCBA has the form (dropping the label I):

$$\Sigma^<(E) = C_0\{N_0 G^<(E+\hbar\omega) + (1+N_0) G^<(E-\hbar\omega)\} \tag{4}$$

$$\Sigma^R(E) = C_0\{N_0 G^R(E+\hbar\omega) + (1+N_0) G^R(E-\hbar\omega)\} + \Lambda(E) \tag{5}$$

$$\Lambda(E) = \tfrac{1}{2} C_0\{G^<(E-\hbar\omega) - G^<(E+\hbar\omega)\} - \tfrac{i}{2} C_0 \frac{P}{\pi}\int dE'\{\frac{G^<(E-E')}{E'+\hbar\omega} - \frac{G^<(E-E')}{E'-\hbar\omega}\} \tag{6}$$

The coupling term $\Lambda(E)$ vanishes for elastic scattering. Typically, the self-energies are initialised to zero and the SCBA equations are iterated until convergence, usually ~ 100 iterations.

The spectral density operator and retarded self-energy may be expressed as:

$$A(E) = \delta(E-H) \to -\tfrac{1}{\pi}\operatorname{Im} G^R(E); \quad \Sigma^R(E) = \Delta^R(E) - i\Gamma^R(E) \tag{7}$$

$\Delta^R(E)$, $\Gamma^R(E)$ are the level shift and level broadening operators respectively. It is well-known that $A(E)$ satisfies the Spectral Density Sum Rule and its trace gives $\rho(E)$ the density of states in energy:

$$\int A(\mathbf{p},E)\,d\mathbf{p} = 1 = \int A(\mathbf{r},E)\,d\mathbf{r} \tag{8}$$

$$\rho(E) = trace\ A(E) \tag{9}$$

$A(\mathbf{p},E)$ *and* $A(\mathbf{r},E)$ are diagonal matrix elements in the momentum and position representations.

In the time domain, the retarded Green function $G^R(t,t')$ has the "causal" response property:

$$G^R(t,t') = 0 \quad for\ t < t' \tag{10}$$

Using the transformation: $\tau = (t-t')$; $T = \tfrac{1}{2}(t+t')$ the Green function $G^R(E)$ in the energy domain E is recovered by Fourier transforming over τ and noting that G^R is independent of T in the stationary limit. The Titchmarsh theorem shows that $G^R(E)$ is a causal function i.e an analytic function in the upper half complex energy plane satisfying a Kramers-Kronig relation or Hilbert Transform. Similarly, $\Sigma^R(E)$ is often *assumed* to be a causal function satisfying the dispersion relation:

$$\Delta^R(E) = \frac{P}{\pi}\int_{-\infty}^{\infty} \frac{\Gamma^R(E')}{E-E'} dE' \tag{11}$$

$\Sigma^R(E)$ need not be a simple causal transform if there are bound states or resonances in the electron spectrum. It may be shown [6] that if $\Sigma^R(E)$ or $\Sigma^R(E)G^R(E)$ is a causal transform then the spectral sum rule (8) is automatically satisfied. From eqns (8-9) it follows that any violations of causality in $\Sigma^R(E)$ or $\Sigma^R(E)G^R(E)$ gives rise to errors in the DOS.

Journal of Physics: Conference Series **526** (2014) 012001 doi:10.1088/1742-6596/526/1/012001

3. Results

We first consider a simple 1D model for the causal retarded self-energy for scattering via optical phonon emission using an exaggerated optical deformation potential C to enhance the details. We choose $\Gamma^R(E) \equiv \Gamma_1(E) \propto \rho_0(E-\hbar\omega) \propto (E-\hbar\omega)^{-1/2}$. The inset to Fig. 3 shows $\Gamma_1(E)$ and $\Delta_1(E)$ obtained from the Hilbert transform (11): $\Sigma_1(E) = \Delta_1(E) - i\Gamma_1(E)$ is fully causal by construction: sum rule satisfied; it is the first Born approximation. Fig. 3 also compares the free DOS $\rho_0(E)$ (purple line) with the perturbed DOS $\rho_1(E)$ (red dashed curve) derived from (7) and (9). The effect of neglecting $\Delta_1(E)$ i.e setting $\Sigma_1(E) \approx -i\Gamma_1(E)$ is shown as the blue curve: this DOS has serious errors due causality violation and sum rule failure. Continuing the iteration leads to the second Born approximation shown in Fig. 4. The fully causal DOS (dashed curve) shows a threshold shift (spectral re-normalisation) compared to the acausal approximation (blue curve).

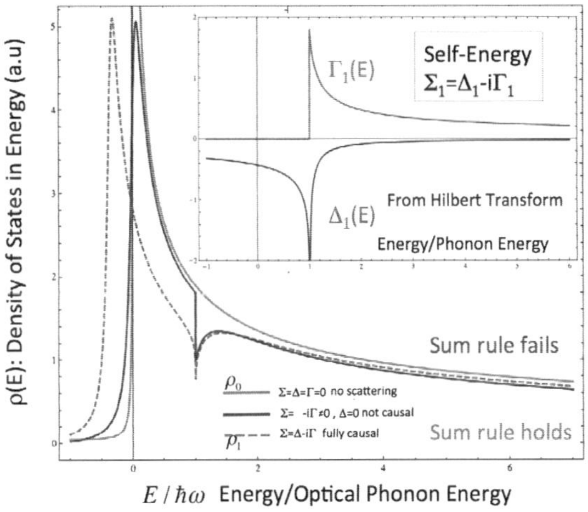

Figure 3: Density of states in the first self-consistent Born approximation for a uniform 1D system.

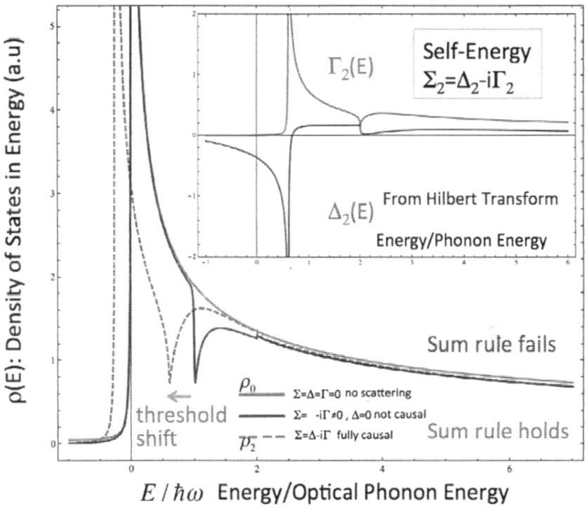

Figure 4: Density of states in second self-consistent Born approximation for a uniform 1D system.

These features are borne out by full NEGF simulations for a long (38 nm) narrow (2.2 nm wide) 3D GAA NWFET silicon device at low gate voltage using the same exaggerated optical deformation potential C=930 eV/nm for just one phonon mode (*g3* phonon: see fig. 5; energy 62 meV) but including absorption processes and using a converged SCBA (~ 100 iterations). Fig. 6 shows results for the DOS in the centre of the channel at low field corresponding closely to the conditions of the simple analytical model. The causal DOS (dashed red curve) and acausal DOS (blue curve) are very similar to the simple model of fig. 4: the kink in the causal DOS shows a threshold shift and the sum rule is satisfied in contrast to the acausal approximation where $\Sigma(E) \approx -i\Gamma(E)$.

Finally, Figures (7- 8) show the energy-resolved current density in the middle of the channel computed from a full 3D self-consistent NEGF simulation for the realistic silicon GAA NW FET device of Figure 1. Acoustic phonon scattering and all the f and g inelastic inter-valley phonon scattering (Figure 5) are included via the Self-Consistent Born Approximation (convergent at around 100 iterations). Deformation potentials and other transport data are from [5]: the optical deformation potential for g phonons is C = 310 eV/nm. Results are shown for fully causal self-energies and for comparison we show results for non-causal self-energies obtained by neglecting the real part of the retarded self-energy. The non-causal results show strong violation of the spectral sum rules and significant departures from the causal spectral current density of states especially for the low gate voltage case (Fig 7: V_G=0.3 V); the effects persist to high gate voltages as shown in Fig. 8 (V_G=0.7 V).

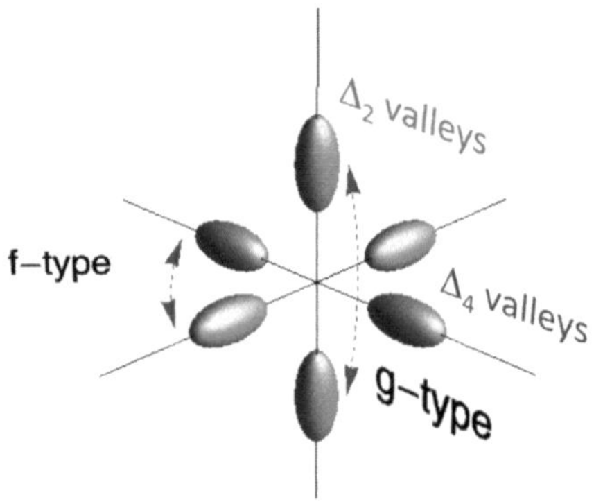

Figure 5: Optical phonon transitions in silicon: *f* and *g* inter-valley phonons.

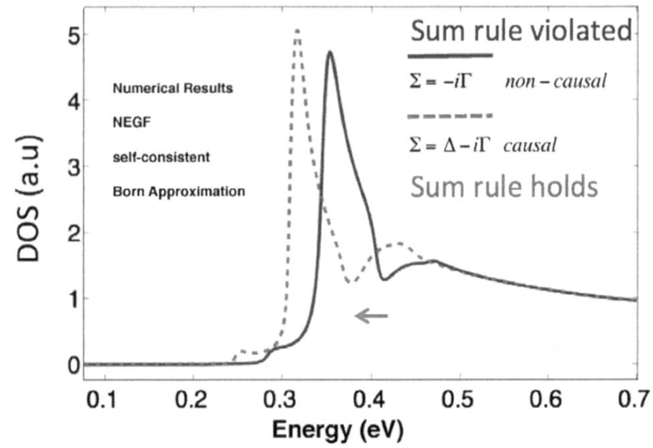

Figure 6: Density of States in middle of channel of a GAA NW FET in full NEGF simulation using frozen field.

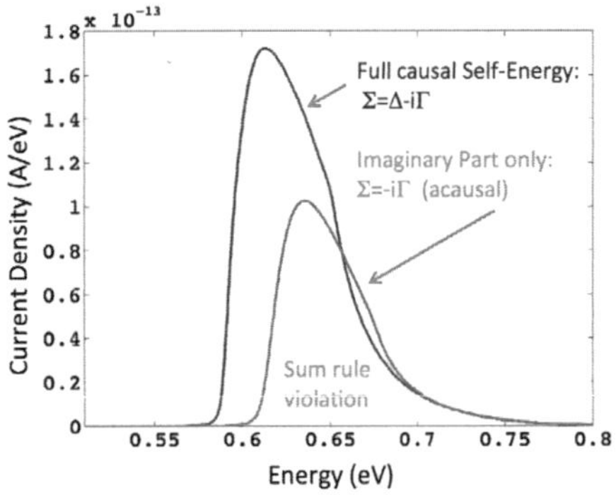

Figure 7: Spectral current density in the middle of the channel at low gate voltage: V_G=0.3 V.

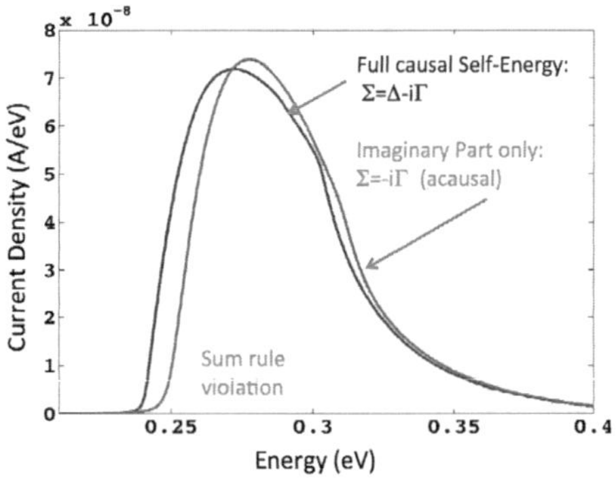

Figure 8: Spectral current density in the middle of the channel at high gate voltage: V_G=0.7 V.

4. Conclusions

Electron-phonon scattering increases with inverse cross-sectional area for silicon nanowire devices with channel diameters < 4 nm. Causally valid inelastic self-energy models are required for the NEGF modelling of nanowires. The spectral density sum rule holds automatically if Σ^R or $\Sigma^R G^R$ are strict causal transforms. A simple 1D model for optical phonon emission shows that neglecting the real part of the self-energy leads to: causality violations, failure of the spectral sum rule and erroneous DOS. The quasi-1D nature of transport in narrow nanowires enhances these effects. The results are confirmed for NEGF computations of realistic dissipative silicon nanowire devices where non-causal approximations are shown to give large errors in the current spectral density.

References

[1] H. Yan, et al, *Nature* **470** 240 (2011).
[2] A. Martinez, J.R. Barker, et al, *J. Computational and Theoretical Nanoscience* **5**, 2289 (2008).
[3] S. Jin, Y. J. Park and H.S. Min, *J. Appl. Phys.* **99** 123719 (2006).
[4] M. Luisier, *App. Phy.s Lett.* **98** 032111 (2011).
[5] M. Aldegunde, A. Martinez and J. R. Barker, *J. Appl. Phys.* **113** 014501 (2013).
[6] J. R. Barker, unpublished.

Journal of Physics: Conference Series **526** (2014) 012002 doi:10.1088/1742-6596/526/1/012002

Description of polarons in ceria using Density Functional Theory.

Christopher WM Castleton and Amy L. Lee
School of Science and Technology, Nottingham Trent University, NG11 8NS, UK
E-mail: Christopher.Castleton@ntu.ac.uk

Jolla Kullgren and Kersti Hermansson
Department of Chemistry, Uppsala University, Box 538, Uppsala, Sweden

Abstract. The performance of various density functional theory (DFT) functionals in reproducing the localization of Ce*4f* electrons to form polarons in cerium dioxide (ceria) is studied. It is found that LDA+U with U=6 eV provides the best description, followed by GGA+U with U=5 eV. Hybrids perform worse, with PBE0 better than HSE06 and HSE03. It is also demonstrated that the improvement in the description of the polarons obtained from LDA+U and GGA+U is due primarily to the effect the U has on the *filled* Ce*4f* states, but the improvement obtained using the hybrids is primarily due to their effect on the *empty* states. This difference can be expected to strongly impact some detailed predictions for the properties of ceria obtained using the two classes of functional.

1. Introduction

The wide bandgap material cerium dioxide (ceria, CeO_2,) has applications in fuel cells, gas sensors and catalysis, but has proved a challenge for theoretical studies. It has an O*2p* based valence band and a nominal Ce*5d* based conduction band, see Fig 1, but in between lies a narrow, empty band of Ce*4f* states. Any potential conduction electron added via doping or defect creation enters this band and localizes onto a single Ce site, breaking the crystal's translational symmetry and inducing a local lattice distortion around the Ce site in question. This forms a "polaron", ie a self-trapped electron, $\mathrm{Ce}_{\mathrm{Ce}}^{-1}$, with an energy gain of roughly 1eV. Polarons generally form coupled to other defects, such as Rh adatoms [2] or oxygen vacancies. Oxygen ions have a formal charge of -2, so removing a neutral oxygen (as part of an oxygen molecule) necessarily leaves behind a +2 charged vacancy $\mathrm{V}_{\mathrm{O}}^{+2}$and two electrons, which in ceria both form polarons. These three remain electrostatically bound in a neutral $\left[\mathrm{V}_{\mathrm{O}}^{+2} - 2\mathrm{Ce}_{\mathrm{Ce}}^{-1}\right]$ complex, which can adopt a wide range of configurations [5]. Conductivity studies [3] have shown that these polarons are well described by the "Small Polaron Model" of Holstein [4], meaning that the radius of the localized charge and lattice distortion are on the order of one lattice spacing. This is also supported by scanning tunnelling microscopy (STM) images of vacancies and vacancy clusters [1], which are best understood if the two electrons are indeed localized in Ce*4f* states on specific Ce ions.

Obtaining a localized description of these $\mathrm{Ce}_{\mathrm{Ce}}^{-1}$ polarons using Density Functional Theory (DFT) is hard, requiring the use of either LDA+U, GGA+U, [7, 8, 9] or hybrid functionals

Journal of Physics: Conference Series **526** (2014) 012002 doi:10.1088/1742-6596/526/1/012002

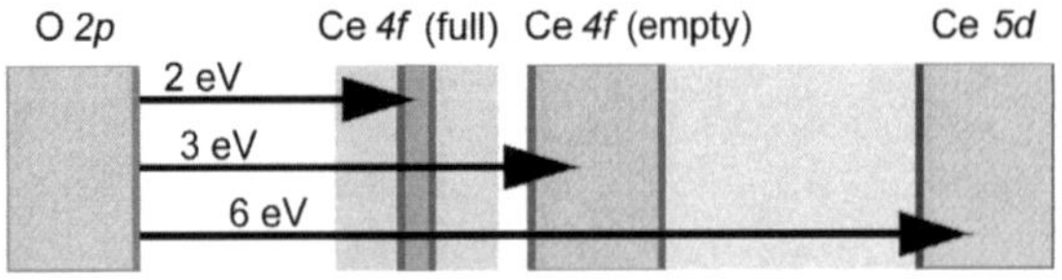

Figure 1. Schematic band diagram for ceria, showing the O*2p* based valence band, and empty Ce*5d* and Ce*4f* bands, plus the filled, localized (polaronic) Ce*4f* states.

[6, 9]. The performance of these in describing the bulk properties (structures, bandgaps etc) of ceria and the sesquioxide, Ce_2O_3, as well as $\left[V_O^{+2} - 2Ce_{Ce}^{-1}\right]$ complexes has been assessed [6, 7, 8, 9]. The best description is obtained using LDA+U with $U = 6$ eV. The hybrids produce qualitatively reasonable descriptions, but are not as good quantitatively. For example, B3LYP does reasonably well for the electronic properties but not for the structure, while PBE0 does well for the structure but not for the bandgaps.

While numerous DFT studies have examined polarons associated with $\left[V_O^{+2} - 2Ce_{Ce}^{-1}\right]$ and other defects, only a few have considered the properties of isolated polarons [10], and we are not aware of any assessment of the quality of the description of lone polarons obtained using different functionals. So we will here compare the description of lone polaron states in ceria using the functionals LDA+U, GGA+U, PBE0 [11] and HSE (both HSE03 [12] and HSE06 [13]).

2. Method

We use plane-wave *ab initio* DFT with the Projector Augmented Wave method [14] (PAW) within the VASP code [15]. $O2s^22p^4$ and $Ce4f^15s^25p^65d^16s^2$ are treated as valence, with a plane-wave cutoff of 300 eV. We use a 96 atom supercell, which is a 2×2×2 multiple of the 12 atom crystallographic cell. K-point integrations use a 4×4×4 and 2×2×2 Monkhorst-Pack grids [16] for LDA/GGA+U and hybrid calculations respectively. We use the rotationally invariant form of LDA+U due to Dudarev et al. [17], together with LDA and with the PBE form of GGA.

3. Results

3.1. Real space charge localization with different functionals

In line with previous studies of $\left[V_O^{+2} - 2Ce_{Ce}^{-1}\right]$ complexes, we find that all the functionals studied can localize Ce*4f* electrons to form small polarons, as illustrated by the isocharge surfaces in Fig 2. These show the partial charge density of the single filled Ce*4f* state. For each functional, surfaces are shown containing 85%, 90% and 95% of the total polaron charge. Clearly, the degree of localization varies, with LDA+U producing the most localized description, closely followed by GGA+U, then the hyrbids PBE0 HSE06 and HSE03. Since the experimental evidence points to very tightly localized Ce*4f* polarons, LDA+U=6 eV seems to be the most successful functional.

In order to check the choice of U=6 eV for LDA+U, we show in Fig 3 the percentage of the supercell contained within the 90% and 95% isocharge surfaces, as a function of U. Clearly, the most localized description is indeed at U=6 eV. Fig 3 also shows the equivalent data (from [7]) for the $\left[V_O^{+2} - 2Ce_{Ce}^{-1}\right]$ complex. Although equally localized near the optimal U, the lone polaron *starts* to localize at slightly higher U values than the vacancy bound polarons. It is apparently easier to localize Ce*4f* electrons when the presence of something else (such as a vacancy) has already broken the translational symmetry of the lattice.

3.2. Calculated band energies

Fig 4 shows band energies taken from the density of states (DOS) of a supercell containing a single Ce*4f* electron, and no other defects. (c.f. schematic in Fig 1). Fig 4 (a) plots the variation of these energies with U for LDA+U. (The inset shows very similar results for GGA+U.) Once localized, the polaron state descends roughly linearly with U, falling 2.1 eV to reach the valence

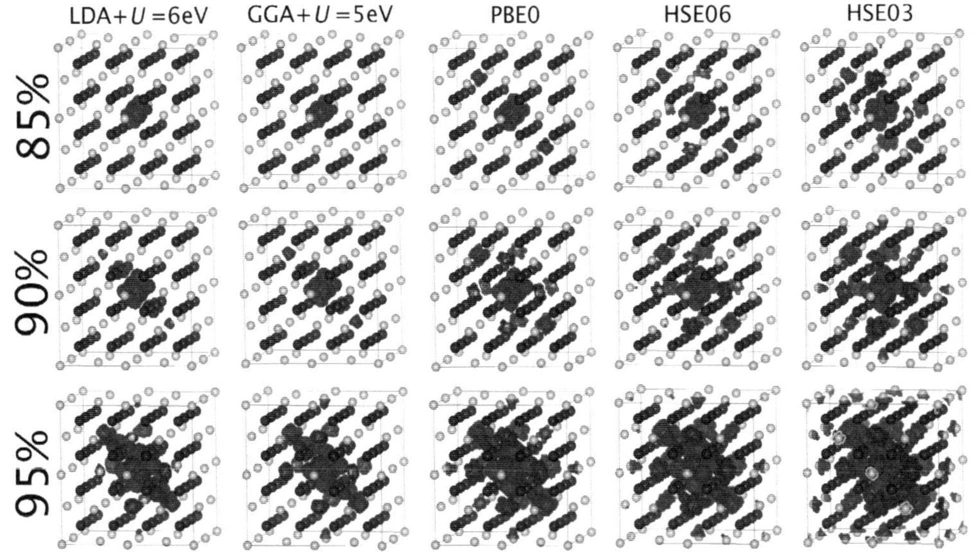

Figure 2. Isocharge surfaces containing 85%, 90% and 95% of the Ce*4f* charge, obtained using the functionals LDA+U=6 eV, GGA+U=5 eV, PBE0, HSE06 and HSE03.

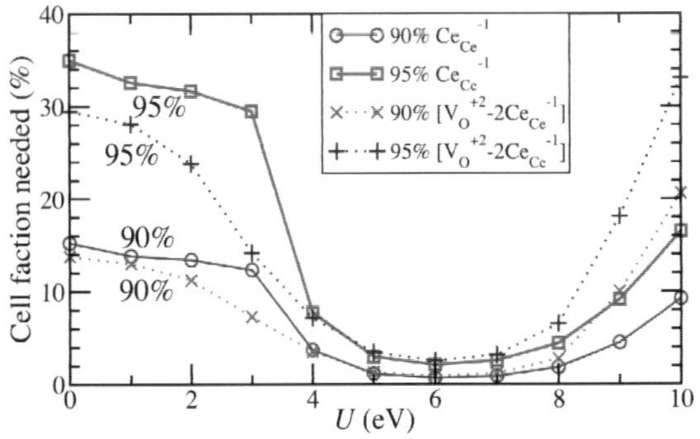

Figure 3. Volume of the supercell contained within the 90% and 95% isosurfaces of Fig 2 for the polaron (solid lines) and for the $\left[V_O^{+2} - 2Ce_{Ce}^{-1}\right]$ complex (dotted lines), as a function of U for LDA+U. Data for $\left[V_O^{+2} - 2Ce_{Ce}^{-1}\right]$ taken from [7].

band at $U \simeq$10 eV. The effect of U on the empty states is much milder, with the Ce*4f*(empty) states rising by only 0.7 eV and the *Ce5d* states falling by 0.4 eV over the same range of U values. In other words, the improvement in the description of the polarons obtained from LDA+U and GGA+U is due primarily to the effect U has on the *filled* Ce*4f* states.

The HSE functional contains a screening range separation parameter μ. Choosing μ=0 gives the PBE0 functional, μ=0.2 gives HSE06, μ=0.3 gives HSE03 and the limit $\mu \to \infty$ recovers pure PBE. In Fig 4 (b) we show the evolution of the band positions with μ. In contrast to LDA/GGA+U, these hybrids have virtually no direct effect on the energy of the filled Ce*4f* state, at least once it has left the Ce*4f* (empty) band. Instead, the improvements are obtained by moving the *empty* states *upwards* in energy. While this is to be expected, given the nature of the hybrids, the difference between this and LDA/GGA+U is striking. Both approaches produce qualitatively reasonable results, but their detailed predictions for the properties of polarons and other features of ceria can therefore be expected to be very different indeed.

Conclusions

We have shown that, while they all correctly localize Ce*4f* electrons in ceria to form small polarons, the functionals LDA+U and GGA+U form much more tightly localized polarons than

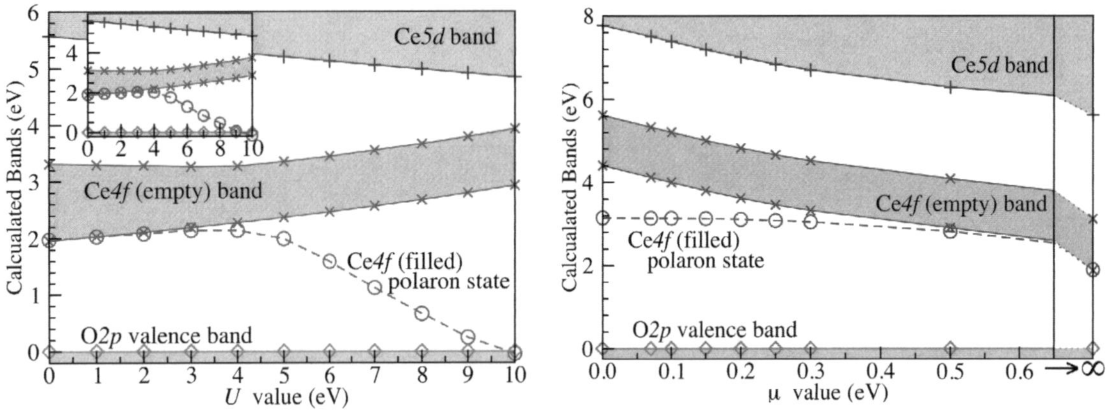

Figure 4. Bands in ceria with one filled Ce4f state, as a function of (a) U for LDA+U (inset: GGA+U), and (b) μ for HSE-type functionals: PBE0 ($\mu = 0$) HSE06 ($\mu = 0.2$), HSE03 ($\mu = 0.3$) and pure PBE ($\mu \to \infty$). Energies plotted relative to the valence band edge.

the hybrids PBE0, HSE06 and HSE03, and are thus more in keeping with the experimental data. We have confirmed that U=6 eV is the optimal value for describing lone polarons with LDA+U. We have also shown that the improvement in the description of polarons obtained from LDA+U and GGA+U is due primarily to the effect that U has on the *filled* Ce4f states, but the improvement obtained using hybrids is due primarily to their effect on the *empty* states. This difference will impact many of the detailed predictions made using the two classes of functional, and this fundamental difference in behaviour has not perhaps received the attention in the literature that it deserves.

References

[1] Esch F, Fabris S, Zhou, Montini T, Africh C, Fornasiero P, Comelli G and Rosei R, 2005 *Science* **309** 752
[2] Lu Z, Yang Z, Hermansson K and Castleton C W M 2014 *J. Mater. Chem. A* **2** 2333
[3] Tuller H L and Nowick A S 1977 *J. Phys. Chem. Solids* **38** 859
Naik I K and Tien T Y 1978 *J. Phys. Chem. Solids* **39** 311
Chang E K and Blumenthal R N 1988 *J. Sol. Stat. Chem.* **72** 330
[4] Holstein T 1959 *Ann. Phys.* **8** 325, **8** 343
[5] Kullgren J, Hermansson K, Castleton C W M, 2012 *J. Chem. Phys.* **137** 044705
[6] Hay P J, Martin R L, Uddin J and Scuseria G E 2006 *J. Chem. Phys.* **125**, 034712
Kullgren J, Castleton C W M, Müller C, Muñoz Ramo D and Hermansson K 2010 *J. Chem. Phys.* **132** 054110
[7] Castleton C W M, Kullgren J and Hermansson K 2007 *J. Chem. Phys.* **127** 244704
[8] Loschen C, Carrasco J, Neyman K N, and Illas F 2007 *Phys. Rev. B* **75** 035115
Andersson D A, Simak S I, Johansson B, Abrikosov I A, and Skorodumova N V 2007 *Phys. Rev. B* **75** 135109
[9] Da Silva J L F Ganduglia-Pirovano M V, Sauer J, Bayer V and Kresse G 2007 *Phys. Rev. B* **75** 045121
[10] Nakayama M, Ohshima H, Nogami M and Martin M 2012 *Phys. Chem. Chem. Phys.* **14** 6079
Zacherle T, Schriever A, De Souza R A and Martin M 2013 *Phys. Rev. B* **87** 134104
Plata J J, Marquez A M, and Sanz J F. 2013 *J. Phys. Chem. C*, 2013 **117** 14502
Plata J J, Marquez A M, and Sanz J F. 2013 *J. Phys. Chem. C*, 2013 **117** 25497
[11] Adamo C and Barone V 1999 *J. Chem. Phys.* **110** 6158
[12] Heyd J, Scuseria G E and Ernzerhof M 2003 *J. Chem. Phys.* **118** 8207
[13] Heyd J, Scuseria G E and Ernzerhof M 2006 *J. Chem. Phys.* **124** 219906
[14] Blöchl P E, Phys. 1994 *Phys. Rev. B* **50** 17953
Kresse G and Joubert D 1999 *Phys. Rev. B* **59** 1758
[15] Kresse G and Furthmüller J 1996 *Comp. Mat. Sci.* **6** 15
[16] Monkhorst H and Pack P 1976 *Phys. Rev. B* **13** 5188
[17] Dudarev S L, Botton G A, Savrasov S Y, Humphreys C J and Sutton A P 1998 *Phys. Rev. B* **57** 1505

Journal of Physics: Conference Series **526** (2014) 012003 doi:10.1088/1742-6596/526/1/012003

Analysis of energy gap opening in graphene oxide

Mark Lundie[1], Željko Šljivančanin[2] and Stanko Tomić[1]

[1] Joule Physics Laboratory, University of Salford, UK
[2] Vinća Institute of Nuclear Sciences, University of Belgrade, Serbia

E-mail: s.tomic@salford.ac.uk

Abstract. The utilisation of graphene structures as photonics materials mandates that an optically active electronic energy gap be formed. Opening of a gap in graphene has been demonstrated by functionalisation with H, F, or O atoms, while experimental observations of graphene oxide have hinted at interesting optical properties, with the potential for absorption of visible light. As such, our analysis is focused on O functionalisation of graphene. We present results from extensive *ab initio* and hybrid DFT calculations, demonstrating the creation of an optically active gap.

1. Introduction

Graphene, owing to its many unique and novel properties, has drawn considerable attention at the prospect that it can be applied to composite materials and next generation nanoelectronics and photonics devices [1]. Unfortunately, the lack of electronic energy gap presents an obstacle to the latter of these applications [2]. Efforts to open a gap in graphene based materials have thus been the subject of much research in recent years.

The electronic structure of pristine graphene in the vicinity of the Fermi level is driven by a 2D network of sp^2-hybridised π bonds. Disruption of this network by surface adsorption of suitable atoms, among other methods, can be achieved, potentially leading to the formation of an energy gap. This has been shown theoretically and validated experimentally for graphene functionalised with atomic hydrogen [3–5], flourine [6], and oxygen [7, 8].

Experimental observations of graphene oxide (GO) have highlighted some interesting optical properties [9]. For example, in aqueous suspension, measurements reveal strong optical absorption in the ultraviolet (UV) [10]. Of key interest is the observation that absorption peaks are shifted toward the visible range with the reduction of the level of oxidation [11].

We investigate the optical properties of epoxy-functionalised graphene by means of *ab intio* and hybrid density functional theory (DFT) calculations. Our results show that this functionalisation opens a wide and optically active band gap.

2. Methodology

Structural calculations are performed using DFT at the generalised gradient approximation (GGA) level of theory, with the well-known PBE functional. These are initially carried out using the GPAW [12] code, an implementation of the projector augmented wave (PAW) method, and subsequently repeated with a plane wave (PW) basis set using CASTEP [13]. In order to make quantitative prediction of electronic and optical properties, the hybrid PBE0 and B3LYP functionals are used for the calculation of these. To make this practical, as well as to allow the use of an all-electron basis set, the CRYSTAL [14, 15] code, in which Bloch functions are constructed from localised Gaussian type orbitals (GTOs), is used for the majority of this study.

To find a suitable GTO basis set, the geometry of clean and partially oxidised graphene cells is relaxed and the bond lengths and binding energies compared to those calculated using PAW and PW basis sets. The most suitable is found to be the Dunning correlation-consistent cc-pCVDZ set [16]. For carbon, adjustment of the most diffuse orbital is required, while for oxygen, the original set is used without modification. Bond lengths are found to be within 0.1 Å and binding energies within 0.01 eV of calculations from the other methods.

Calculations are performed using a rectangular unit cell containing 8 C atoms and 4 O atoms. The geometry is relaxed and structural parameters calculated using the PBE functional [17]. The relaxed geometry is then used as input for band structure calculations using the PBE0 and B3LYP hybrid functionals [18–20].

3. Results and discussion

Fig. 1(a) shows the most stable configuration of fully oxidised graphene [21]. Each O adatom forms an epoxy functional group with two C atoms, alternating above and below the graphene lattice, thereby minimising Coulomb repulsion.

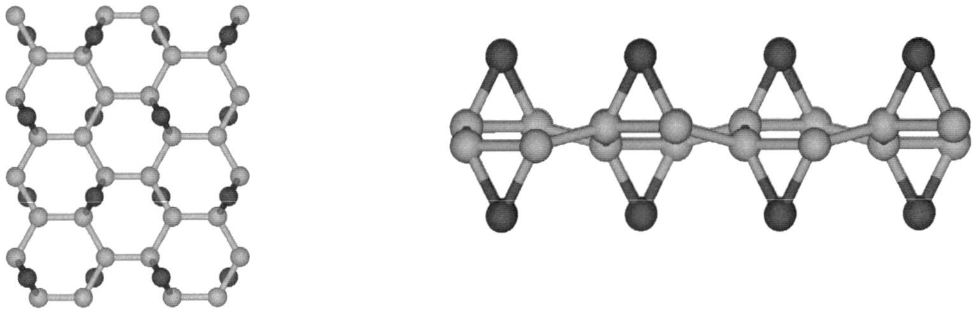

(a) Input. (b) Optimised.

Figure 1. Surface supercell.

The resultant strain from the formation of sp^3-hybridised bonds between the C and O $2p_z$ orbitals causes vertical displacement of the C atoms from their 2D hexagonal lattice sites, as shown in Fig. 1(b). The binding energy per O adatom, calculated relative to the energy of a free O atom, is 3.38 eV, which indicates the structure to be highly stable. The mean C-O bond length is 0.15 Å, while the C atoms are displaced vertically by 0.16 Å relative to their positions in the honeycomb lattice of graphene. The sp^2 to sp^3 bond rehybridisation also causes substantial changes to the band structure, depicted in Fig. 2(a), in comparison to that of pristine graphene. This, along with the total density of states plotted in Fig. 2(b), clearly identify fully oxidised graphene to be an insulator, with a band gap of 6.50 eV, as calculated using B3LYP. As can be expected, the gap predicted by pure *ab initio* theory, using the PBE functional, is much smaller at 4.09 eV. The effective masses along the direction $\Gamma \rightarrow$ M are predicted as $m_e^* = 1.36\ m_0$ and $m_h^* = 0.56\ m_0$, and along $\Gamma \rightarrow$ K as $m_e^* = 0.40\ m_0$ and $m_h^* = 0.44\ m_0$.

Normalised dipole matrix elements are presented in Table 1, showing the fundamental gap to be optically active. Hybrid functionals again predict dramatically different values to those

Journal of Physics: Conference Series **526** (2014) 012003 doi:10.1088/1742-6596/526/1/012003

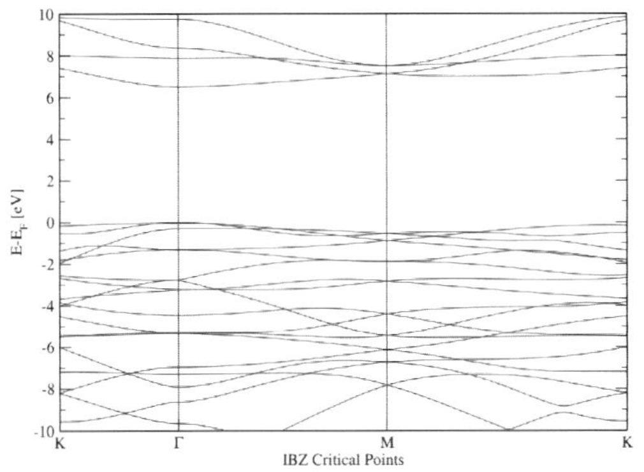

(a) Band structure.

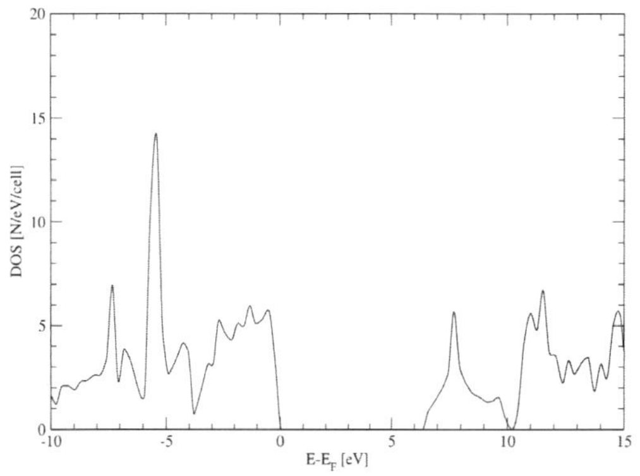

(b) Total density of states.

Figure 2. Band structure and density of states.

from pure *ab initio* theory. These values (from hybrid functionals) are $\sim 1/3$ of those for the direct gap bulk zincblende semiconductors, an indication of strong absorption of UV light. For example, GaAs has dipole matrix elements $p_x = p_y = p_z = 9.69$ eV Å.

Functional	p_x [eV Å]	p_y [eV Å]	p_z [eV Å]
B3LYP	3.66	0.00	0.00
PBE0	3.88	0.00	0.00

Table 1. Dipole matrix elements.

Acknowledgments
The authors acknowledge the Royal Society of London for the grants "Functionalized Graphene for Solar Cell Devices: Insight from Ab Initio Theory" and "High Performance Computing in Modelling of Innovative Photo-Voltaic Devices." The authors also thank Barry Searle & Leonardo Bernasconi for useful discussions. Mark Lundie acknowledges the University of Salford for the award of Graduate Teaching Studentship. Željko Šljivančanin acknowledges financial support from the Serbian Ministry of Education and Science under grant no. 171033 and from the COST Action CM1104.

References

[1] Novoselov K, Geim A, Morozov S, Jiang D, Zhang Y, Dubonos S, Grigorieva I and Firsov A 2004 *Science* **306** 666–669
[2] Bonaccorso F, Sun Z, Hasan T and Ferrari A C 2010 *Nat. Photonics* **4** 611–622
[3] Balog R, Jørgensen B, Wells J, Lægsgaard E, Hofmann P, Besenbacher F and Hornekr L 2009 *J. Am. Chem. Soc.* **131** 8744–8745
[4] Balog R, Jørgensen B, Nilsson L, Andersen M, Rienks E, Bianchi M, Fanetti M, Laegsgaard E, Baraldi A, Lizzit S, Šljivancanin Ž, Besenbacher F, Hammer B, Pedersen T G, Hofmann P and Hornekaer L 2010 *Nat. Mater.* **9** 315–319
[5] Balog R, Andersen M, Jørgensen B, Šljivancanin Ž, Hammer B, Baraldi A, Larciprete R, Hofmann P, Hornekr L and Lizzit S 2013 *ACS Nano* **7** 3823–3832
[6] Nair R R, Ren W, Jalil R, Riaz I, Kravets V G, Britnell L, Blake P, Schedin F, Mayorov A S, Yuan S, Katsnelson M I, Cheng H M, Strupinski W, Bulusheva L G, Okotrub A V, Grigorieva I V, Grigorenko A N, Novoselov K S and Geim A K 2010 *Small* **6** 2877–2884

Journal of Physics: Conference Series **526** (2014) 012003 doi:10.1088/1742-6596/526/1/012003

[7] Vinogradov N A, Schulte K, Ng M L, Mikkelsen A, Lundgren E, Martensson N and Preobrajenski A B 2011 *J. Phys. Chem. C* **115** 9568–9577
[8] Larciprete R, Fabris S, Sun T, Lacovig P, Baraldi A and Lizzit S 2011 *J. Am. Chem. Soc.* **133** 17315–17321
[9] Niyogi S, Bekyarova E, Itkis M, McWilliams J, Hamon M and Haddon R 2006 *J. Am. Chem. Soc.* **128** 7720–7721
[10] Eda G, Lin Y Y, Mattevi C, Yamaguchi H, Chen H A, Chen I S, Chen C W and Chhowalla M 2010 *Adv. Mater.* **22** 505
[11] Kong C Y, Song W L, Meziani M J, Tackett K N, Cao L, Farr A J, Anderson A and Sun Y P 2012 *J. Supercrit. Fluids* **61** 206–211
[12] Enkovaara J, Rostgaard C, Mortensen J J, Chen J, Duak M, Ferrighi L, Gavnholt J, Glinsvad C, Haikola V, Hansen H A, Kristoffersen H H, Kuisma M, Larsen A H, Lehtovaara L, Ljungberg M, Lopez-Acevedo O, Moses P G, Ojanen J, Olsen T, Petzold V, Romero N A, Stausholm-Mller J, Strange M, Tritsaris G A, Vanin M, Walter M, Hammer B, Hkkinen H, Madsen G K H, Nieminen R M, Nrskov J K, Puska M, Rantala T T, Schitz J, Thygesen K S and Jacobsen K W 2010 *J. Phys.: Condens. Matter* **22** 253202
[13] Clark S J, Segall M D, Pickard C J, Hasnip P J, Probert M J, Refson K and Payne M C 2005 *Z. Kristallogr.* **220** 567–570
[14] Bush I J, Tomić S, Searle B G, Mallia G, Bailey C L, Montanari B, Bernasconi L, Carr J M and Harrison N M 2011 *Proc. R. Soc. A* **467** 2112–2126
[15] Dovesi R, Saunders V R, Roetti C, Orlando R, Zicovich-Wilson C, Pascale F, Civalleri B, Doll K, Harrison N M, Bush I J, D'Arco P and Llunell M 2009 *CRYSTAL 09 Manual*
[16] Dunning T H 1989 *J. Chem. Phys.* **90** 1007–1023
[17] Perdew J P 1985 *Int. J. Quantum Chem.* **28** 497–523
[18] Perdew J P, Ernzerhof M and Burke K 1996 *J. Chem. Phys.* **105** 9982–9985
[19] Becke A D 1993 *J. Chem. Phys.* **98** 1372–1377
[20] Tomić S, Montanari B and Harrison N M 2008 *Physica E* **40** 2125–2127
[21] Šljivančanin Ž, Milošević A S, Popović Z S and Vukajlović F R 2013 *CARBON* **54** 482–488

Ab initio parameterisation of the 14 band k·p Hamiltonian: Zincblende study

Mark Lundie and Stanko Tomić
Joule Physics Laboratory, University of Salford, UK
E-mail. s.tomic@salford.ac.uk

Abstract.
Despite continued and rapid progress in high performance computing, atomistic level device modelling is still largely out of reach, necessitating the use of quantum mechanical continuum methods, including **k·p** perturbation theory. The effective use of such methods requires reliable parameterisation, often obtained from experiment and *ab initio* calculations. A major limitation of this, the systematic tendency of *ab initio* density functional theory to underestimate semiconducting material energy band gaps and related properties, can be greatly improved upon by the inclusion of exact exchange, calculated within the Hartree-Fock formalism. We demonstrate that the 14 band **k·p** Hamiltonian can be effectively parameterised using this method, at greatly reduced cost in comparison to GW methods.

1. Introduction

The modelling of semiconductor devices typically requires multiscale methods, whereupon calculations made at fundamental levels of theory are used to feed more approximate, but critically less computationally expensive, methods. *Ab initio* density functional theory (DFT) is commonly used for atomistic level calculations, which are then used to parameterise tight-binding (TB), empirical pseudopotential (EPM), or multiband **k·p** Hamiltonians [1]. Kohn-Sham DFT, however, does not yield the real band structure of semiconductors. Although a good approximation, the energy band gaps are systematically underestimated in this class of materials.

Many body perturbation theory, typically Green's function methods, can be used to more accurately predict the band gaps [2], but at considerable computational expense. An efficient and reasonably accurate alternative is the use of hybrid functionals [3, 4], which incorporate into the DFT exchange-correlation energy either a screened long range Coulomb term or a fraction of exact exchange calculated from Hartree-Fock theory using the Kohn-Sham orbitals.

Using this method, we calculate the necessary energy band gaps, dipole matrix elements, and effective masses at characteristic points in the first irreducible Brillouin zone (IBZ), i.e. at Γ point, to parameterise the 8 and 14 band **k·p** Hamiltonians. Although the 8 band **k·p** Hamiltonian, which has C_{4v} symmetry, is popular and widely used, it overlooks the correct C_{2v} atomistic symmetry of the zincblende (ZB) lattice [5]. In order to restore the correct symmetry it is necessary to implement the 14 band model, including couplings induced by the second conduction band (labelled as Γ_{5c} in fig. 1) stemming from *p*-bonding states of atoms in the ZB lattice [5]. We compare both to the DFT band structures in demonstration of this. We further

Journal of Physics: Conference Series **526** (2014) 012004 doi:10.1088/1742-6596/526/1/012004

show the importance of non-locality in hybrid DFT calculations to the accurate prediction of band parameters.

2. Methodology

All DFT calculations are performed using the CRYSTAL [6, 7] code, which implements localised basis sets of Gaussian type orbitals (GTOs). The calculation of exact exchange is very efficient using this basis. The hybrid PBE0 [8] and B3LYP [9] exchange correlation functionals are used, the former incorporating 25% of exact exchange energy to the latter's 20%. The IBZ of the ZB lattice unit cell is sampled using the Monkhorst-Pack scheme [10], with shrinking factors 16×16×16. Dipole matrix elements $\mathbf{p}_{ij}$ are calculated from Bloch functions, $u_{\mathbf{k}i}(\mathbf{r})$, such that:

$$-i\frac{\hbar}{m_0}\mathbf{p}_{ij} = [E_i - E_j]\int d\mathbf{r}\, u^*_{\mathbf{k}i}(\mathbf{r})\,\hat{\mathbf{r}}\,u_{\mathbf{k}j}(\mathbf{r}). \tag{1}$$

From the dipole matrix elements, Eq. 1, the Kane energies are calculated as follows:

$$E_P = \frac{2}{m_0}\left|\langle u_{c,0}|\hat{\mathbf{p}}|u_{v,0}\rangle\right|^2. \tag{2}$$

The *kppw* code [11], parameterised with the *ab initio* data, is used for all **k·p** calculations.

3. Results and discussion

Table 1 contains all the relevant parameters required for the parametrisation of unstrained 8 and 14 band **k·p** Hamiltonians for GaAs and CdSe, two common semiconductors that crystallise in the ZB structure.

Direction	Parameter	GaAs	CdSe
	E_{g_0} [eV]	1.42	2.16
	E_{g_1} [eV]	4.93	7.43
	E_{P_0} [eV]	24.69	17.52
	E_{P_1} [eV]	1.05	0.79
	E_{P_2} [eV]	13.49	13.24
	m^*_c/m_0	0.056	0.114
[001]	m^*_{lh}/m_0	0.056	0.118
[111]	m^*_{lh}/m_0	0.047	0.098
[001]	m^*_{hh}/m_0	0.334	0.597
[111]	m^*_{hh}/m_0	0.776	1.241
[001]	γ_1	10.39	5.07
[111]	γ_1	11.22	5.51
	γ_2	3.70	1.69
	γ_3	4.97	2.35

Table 1: Calculated parameters for 8 and 14 **k·p** Hamiltonians, for GaAs and CdSe ZB lattices obtained with B3LYP functional.

Effective masses are defined by the curvature of the band dispersion at high symmetry points in the IBZ, in this case Γ:

$$\left(\frac{1}{m^*}\right)_{\mu\nu} = \frac{1}{\hbar^2}\frac{\partial^2 E_k}{\partial k_\mu \partial k_\nu} \tag{3}$$

An alternative definition of the effective mass, from **k**·**p** theory, depends on the interaction between bands around the conduction and valence band edges, where i is the band for which the effective mass is being calculated:

$$\frac{1}{m_i^*} = \frac{1}{m_0} + \frac{2}{m_0} \sum_{i \neq j} \frac{|p_{ij}|^2}{|E_i - E_j|} \tag{4}$$

The effective mass calculated using this expression can be systematically improved by considering additional interactions from remote bands. For the methodology to be consistent, convergence of the two should be observed as further bands are considered in Eq. 4. Indeed, this is observed, as shown in Table 2. We use shortened notation: $p_0 = p_{\Gamma_{1c},\Gamma_{5v}}$, $p_1 = p_{\Gamma_{5c},\Gamma_{1c}}$, $p_2 = p_{\Gamma_{5c},\Gamma_{5v}}$.

	Eq. (3)	Eq. (4)		
		p_0	p_0, p_1	p_0, p_1, p_2
GaAs	0.0561	0.0552	0.0561	0.0561

Table 2: Convergence of calculated effective mass at the conduction band edge of GaAs as additional band couplings, p_i, taken into account.

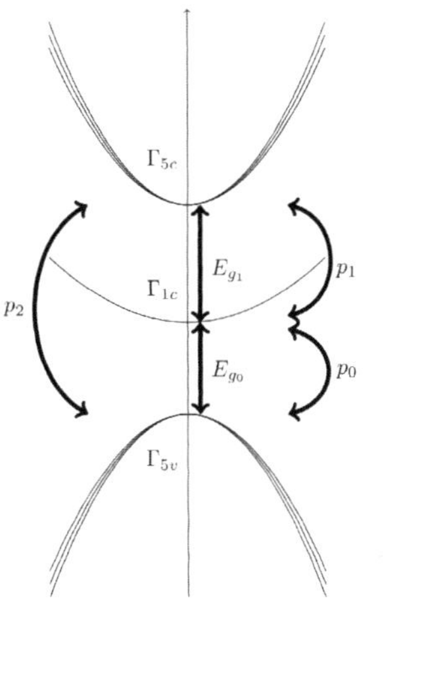

Figure 1: Schematic of ZB band structure around Γ point. Top of valence band, Γ_{5v}, bottom of conduction band, Γ_{1c}, and second conduction band, Γ_{5c}. Energy gaps, E_{g_i}, and coupling parameters, p_i, correspond to the notation in table 1.

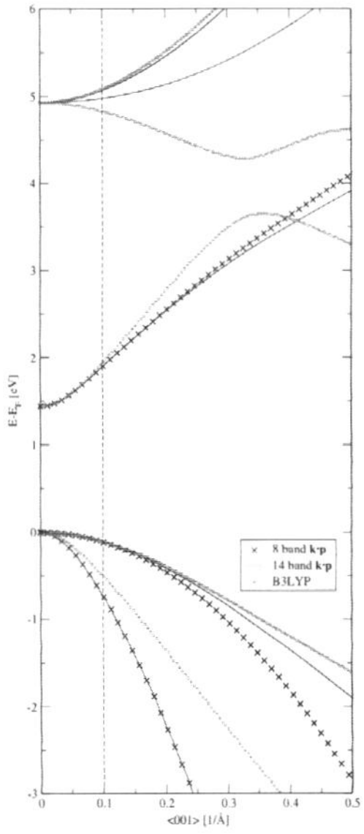

Figure 2: Band structure of GaAs along [001] direction in the IBZ, obtained by DFT-B3LYP (open circles), 8 band **k**·**p** Hamiltonian (crosses), and 14 band **k**·**p** Hamiltonian (lines).

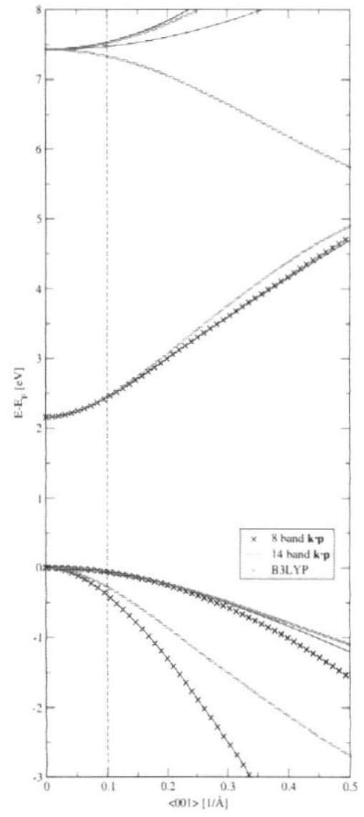

Figure 3: Band structure of CdSe along [001] direction in the IBZ, obtained by DFT-B3LYP (open circles), 8 band **k**·**p** Hamiltonian (crosses), and 14 band **k**·**p** Hamiltonian (lines).

The Luttinger parameters relate to the valence band edge effective masses along directions of high symmetry in the IBZ and largely determine the band curvature. They are calculated from appropriate pairs of the following system of equations, along different high symmetry directions:

$$\begin{aligned} \left(\frac{m_0}{m^*_{\text{hh}}}\right)^{[001]} &= \gamma_1 - 2\gamma_2, & \left(\frac{m_0}{m^*_{\text{hh}}}\right)^{[110]} &= \frac{1}{2}(2\gamma_1 - \gamma_2 - 3\gamma_3), & \left(\frac{m_0}{m^*_{\text{hh}}}\right)^{[111]} &= \gamma_1 - 2\gamma_3, \\ \left(\frac{m_0}{m^*_{\text{lh}}}\right)^{[001]} &= \gamma_1 + 2\gamma_2, & \left(\frac{m_0}{m^*_{\text{lh}}}\right)^{[110]} &= \frac{1}{2}(2\gamma_1 + \gamma_2 + 3\gamma_3), & \left(\frac{m_0}{m^*_{\text{lh}}}\right)^{[111]} &= \gamma_1 + 2\gamma_3, \end{aligned} \quad (5)$$

As the effective mass tensors are anisotropic, we obtain different Luttinger parameters along each direction, as detailed in table 1.

Fig. 1 depicts the schematic band structure of the ZB structure, taken into account by the 14 band **k**·**p** Hamiltonian. The 8 band Hamiltonian (C_{4v}) includes only the Γ_{5v} valence and Γ_{1c} conduction bands, coupled by the p_0 dipole interaction. The p_1 interaction is isomorphic with p_0, however the inclusion of the p_2 interaction between Γ_{5v} and Γ_{5c} bands has the effect of reducing the symmetry to C_{2v}, as required to resemble the correct atomistic structure of ZB lattice.

Figs. 2 and 3 show the band structures along the <001> direction in the IBZ for GaAs and CdSe, respectively. Largely good agreement between the DFT and **k**·**p** calculations is seen up to 0.1 Å^{-1}. In addition to observing the correct C_{2v} symmetry, the 14 band model shows a notable improvement in its depiction of the heavy hole band well beyond this region. Of greatest note is the inclusion of the higher conduction bands in the 14 band model.

4. Conclusion

We have demonstrated that the use of hybrid functionals in DFT calculations can be used to effectively and accurately parameterise multiband **k**·**p** Hamiltonians for some common ZB semiconductors. Further, we have shown that the 14 band model, by inclusion of higher conduction bands and having the correct C_{2v} symmetry of the ZB lattice, affords greater accuracy in the prediction of the band structure than the more widely used 8 band model, which is typically synonymous with **k**·**p** perturbation theory.

Acknowledgments

The authors acknowledge the EPSRC for the grant "Enhanced multiple exciton generation in colloidal quantum dots." The authors also thanks Leonardo Bernasconi, Jacek Miloszewski, and Nenad Vukmirović for useful discussions. Mark Lundie acknowledges the University of Salford for the award of Graduate Teaching Studentship.

References

[1] Christensen N E 1984 *Phys. Rev. B* **30** 5753
[2] Rideau D, Feraille M, Ciampolini L, Minondo M, Tavernier C and Jaouen H 2005 *Phys. Rev. B* **74** 195208
[3] Heyd J, Peralta J E, Scuseria G E and Martin R L 2005 *J. Chem. Phys.* **123** 174101
[4] Tomić S, Montanari B and Harrison N M 2008 *Physica E* **40**
[5] Tomić S and Vukmirović N 2011 *J. Appl. Phys.* **110** 053710
[6] Dovesi R, Saunders V R, Roetti C, Orlando R, Zicovich-Wilson C, Pascale F, Civalleri B, Doll K, Harrison N M, Bush I J, D'Arco P and Llunell M 2009 *CRYSTAL 09 Manual*
[7] Bush I J, Tomić S, Searle B G, Mallia G, Bailey C L, Montanari B, Bernasconi L and Harrison N M 2011 *Proceedings of the Royal Society A* **467** 2112
[8] Perdew J P, Ernzerhof M and Burke K 1996 *J. Chem. Phys.* **105** 9982–9985
[9] Becke A D 1993 *J. Chem. Phys.* **98** 1372–1377
[10] Monkhorst H J and Pack J D 1976 *Phys. Rev. B* **13**(12) 5188–5192
[11] Tomić S, Sunderland A G and Bush I J 2006 *J. Mater. Chem.* **16**(20) 1963–1972

Journal of Physics: Conference Series **526** (2014) 012005 doi:10.1088/1742-6596/526/1/012005

Theoretical studies of excitons in type II CdSe/CdTe quantum dots

Jacek M. Miloszewski and Stanko Tomić
School of Computing, Science & Engineering, University of Salford, United Kingdom

David Binks
School of Physics and Astronomy, University of Manchester, United Kingdom

Abstract. We present a method for calculating exciton and bi-exciton energies in type-II colloidal quantum dots. Our methodology is based on an 8-band $\mathbf{k} \cdot \mathbf{p}$ Hamiltonian of the zinc-blend structure, which incorporates the effects of spin-orbit interaction, strain between the core and the shell and piezoelectric potentials. Exciton states are found using the configuration interaction (CI) method that explicitly includes the effects of Coulomb interaction, as well as exchange and correlation between many-electron configurations. We pay particular attention to accurate modelling of the electrostatic interaction between quasiparticles. The model includes surface polarization and self-polarization effects due to the large difference in dielectric constants at the boundary of the QD.

1. Introduction

In a standard solar cell, all of the energy of an absorbed photon in excess of the effective bandgap of the material is dissipated as heat and essentially wasted. In colloidal QDs (for example made of CdSe, CdTe), this excess photon energy can be utilized due to a process known as multiple exciton generation (MEG) or direct carrier multiplication. In this process the high energy photon creates a high energy exciton that can decay into a bi-exciton (shown in Fig. 1a). For this process to occur the the energy of the exciton (which usually consists of a high energy electron and a hole in the ground state) has to be at least twice as big as the energy of the effective optical gap $E_{e_n} - E_{h_0} \geq 2\,|E_{e_0} - E_{h_0}|$, where e_0 and h_0 are the electron and hole ground states, and e_n is a state higher in the conduction band. This allows for full utilization of high energy photons and dramatically increases solar cell efficiency. The MEG process competes with other radiative and non-radiative recombination processes, most of all with Auger cooling [1, 2, 3]. To further increase the solar cell efficiency it is necessary to optimize the shape and composition of the QD in order to maximize the ratio of MEG to cooling processes.

It is hypothesized that one way to increase the MEG/cooling ratio is to introduce a shell layer. Contrary to a simple single material colloidal QD, a core/shell structure can offer extra degrees of freedom in designing devices with desirable properties. In this paper we focus on the development of a theoretical method that allows for quick and reliable determination of exciton and bi-exciton energies in core/shell structures.

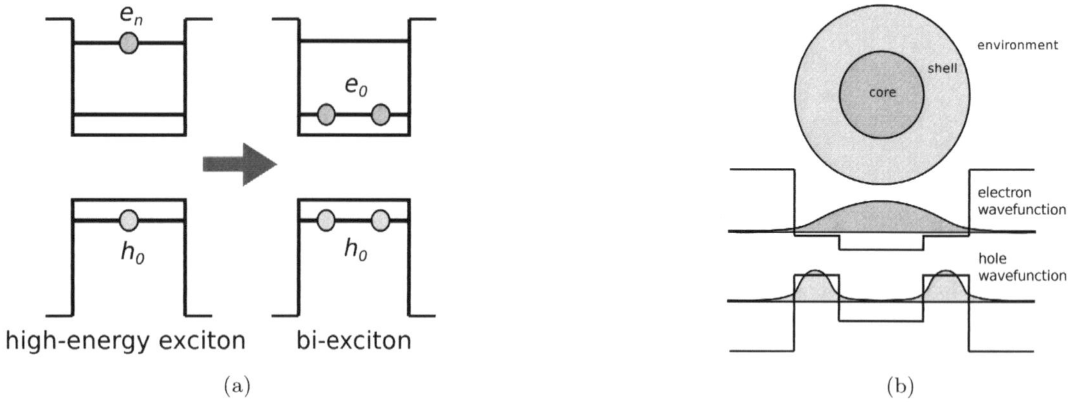

Figure 1: (a) Representation of the multiple exciton generation process. An exciton that consists of high energy electron e_n and ground state hole h_0 decays into a bi-exciton state that consists of 2 ground state electrons e_0 and 2 ground state holes h_0. (b) Structure of a CdSe/CdTe core/shell type-II quantum dot. In CdSe/CdTe core/shell QD hole wavefunctions are contained in the shell region.

2. Theory

Contrary to single material QDs, a core/shell QD can form a type-II structure in which the conduction and valence band states are spatially separated. This separation is schematically shown in Fig. 1b. In the case of QDs with a CdSe core and CdTe shell, the valence band states are usually strongly confined to the shell region while conduction band states are spread equally over the whole structure.

Our approach can be divided into two parts: in the first part we calculate single-particle eigenfunctions and eigenenergies of the confined electrons and holes, while in the second part we use functions from step one to build the many-body Hamiltonian that is used to calculate many-body states such as excitons and bi-excitons.

To find single-particle states we apply an 8-band $\mathbf{k} \cdot \mathbf{p}$ Hamiltonian that includes continuum mechanical strain and piezoelectric effects [4, 5, 6]. The Hamiltonian is parameterized using material parameters taken either from experiment or from ab initio hybrid density functional calculations using B3LYP and PBE0 functionals [7].

Because of the large difference in dielectric permittivities, each charge inside the dot induces significant surface charges which strongly interact with the charges inside the dot. This is known as the surface polarization effect [8, 9]. The exact formula for the Coulomb potential can be obtained by directly solving the Poisson equation, resulting in two terms: $V_{\mathrm{P}}(\mathbf{r}_1, \mathbf{r}_2)$, two-particle interaction term, and $V_{\mathrm{SP}}(\mathbf{r})$, the self-polarization potential originating from the interaction of a charge with its own induced surface charge. The full potential is

$$V(\mathbf{r}_1, \mathbf{r}_2) = V_{\mathrm{P}}(\mathbf{r}_1, \mathbf{r}_2) + V_{\mathrm{SP}}(\mathbf{r}_1) + V_{\mathrm{SP}}(\mathbf{r}_2). \quad (1)$$

To speed up calculations, in this work we use the following approximation for the two-particle interaction

$$V_{\mathrm{P}}(\mathbf{r}_1, \mathbf{r}_2) \approx V_{\mathrm{P}}(|\mathbf{r}_1 - \mathbf{r}_2|) = \frac{e}{4\pi\epsilon_0 |\mathbf{r}_1 - \mathbf{r}_2| \, \epsilon(|\mathbf{r}_1 - \mathbf{r}_2|)}, \quad (2)$$

where $\epsilon(r) = \epsilon_{\mathrm{dot}}$ for $r < R_{\mathrm{b}}$ and $\epsilon(r) = \epsilon_{\mathrm{env}}$ for $r > R_{\mathrm{b}}$. We assumed that the dielectric permittivity is the same in both core and shell and R_{b} is the radius of the whole QD. The

Journal of Physics: Conference Series **526** (2014) 012005 doi:10.1088/1742-6596/526/1/012005

self-polarization term $V_{SP}(\mathbf{r})$ is calculated using the method developed by Bolcatto et al. [10]. The many-body Hamiltonian contains only particle-conserving terms, and is given by

$$\begin{aligned} H &= \sum_i E_i \hat{e}_i^\dagger \hat{e}_i - \sum_i E_i \hat{h}_i^\dagger \hat{h}_i + \frac{1}{2}\sum_{ijkl} V_{iljk} \hat{e}_i^\dagger \hat{e}_j^\dagger \hat{e}_k \hat{e}_l \\ &+ \frac{1}{2}\sum_{ijkl} V_{iljk} \hat{h}_i^\dagger \hat{h}_j^\dagger \hat{h}_k \hat{h}_l - \sum_{ijkl} \left(V_{iljk} - V_{ikjl}\right) \hat{e}_i^\dagger \hat{h}_j^\dagger \hat{h}_k \hat{e}_l, \end{aligned} \tag{3}$$

where Coulomb integrals V_{iljk} are defined as

$$V_{iljk} = \sum_{b=1}^{8}\sum_{b'=1}^{8} \int d^3\mathbf{r} \int d^3\mathbf{r}' \psi_b^{(i)}(\mathbf{r})^* \psi_b^{(l)}(\mathbf{r}) V(\mathbf{r}, \mathbf{r}') \psi_{b'}^{(j)}(\mathbf{r}')^* \psi_{b'}^{(k)}(\mathbf{r}'). \tag{4}$$

In Eq. (3) we ignore terms that do not conserve the number of particles so that we can solve it for excitons and bi-excitons separately. To get excitonic states we write the Hamiltonian in a two particle basis (one electron and one hole) $|i,j\rangle = |e_i\rangle |h_j\rangle$, and for bi-excitons we use a four particle basis $|i,j,k,l\rangle = |e_i\rangle |h_j\rangle |e_k\rangle |h_l\rangle$.

3. Results

We apply our method to study CdSe/CdTe quantum dots with varying shell thickness. The CdSe core has radius of 15Å. Because the difference between the dielectric constants of CdSe and CdTe is very small, they are assumed to be equal, with $\epsilon_{\text{dot}} = 5.8$. The dielectric constant of the environment is assumed to be $\epsilon_{\text{env}} = 1.1$. In the results presented here the effect of self-polarization is ignored, i.e., the formula for Coulomb potential is limited to Eq. (2). For CI calculations we use the first 12 electron and 12 hole states (always including all degenerate states).

To evaluate the the effect of the many-electron interaction in CdSe/CdTe QDs we compare the difference between the ground state exciton energy and the energy difference between e_0 and h_0 single particle states. From Fig. 2a one can see that the exciton binding energy is very large, up to 450 meV ($\sim$ 24% of the bulk bandgap) for a pure CdSe QD (corresponds to the zero thickness point on the graph). It decreases with the addition of CdTe shell, but even for 15Å thickness, it is still around 200 meV, which is significant. This suggest that the Coulomb potential is crucial in accurate determination of energy levels.

To estimate the importance of polarization, we compare exciton and bi-exciton energies calculated with and without the polarization effect. The lack of polarization is implemented by setting $\epsilon_{\text{env}} = \epsilon_{\text{dot}} = 5.8$. Fig. 2b indicates that polarization also plays a very important role in exciton and bi-exciton energy. After taking into consideration the polarization effect, the exciton energy is reduced by roughly 100 meV (200 meV in case of bi-excitons).

Finally, we investigate the influence of exchange and correlation. In Fig. 2c we compare full CI calculations obtained by diagonalization of the many-electron Hamiltonian in Eq. (3) with calculation that do not take into account the correlation effect. To build the many-body Hamiltonian we use the first 24 single particle states (12 hole and 12 electron states) which is enough to get a reasonable convergence of the first exciton and bi-exciton states.

To calculate many-body states that do not include the correlation effect we use only single particle ground states (2 electron and 4 hole) in the construction of the many-body Hamiltonian. We found that the correlation interaction has negligible influence on exciton energies for both core and core/shell structures. It does however have a strong impact on bi-exciton energies, especially in case of type-II structures. Figure 2c shows a comparison of full CI calculations with the corresponding Hartree approximation equivalent (no correlation). The difference ranges from 20 meV for type-I QDs to 60 meV for type-II structures with 15Å shell.

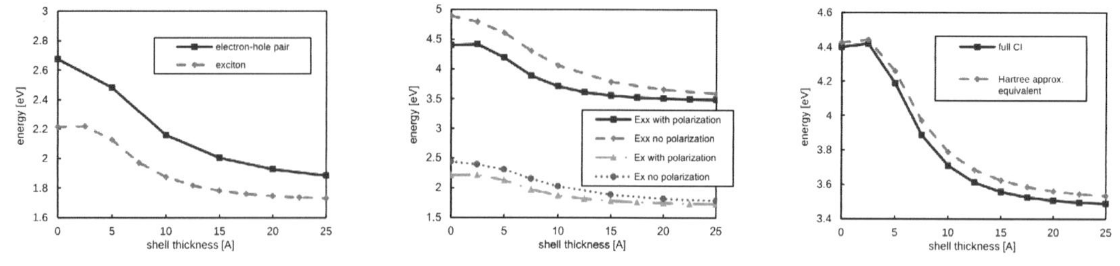

(a) (b) (c)

Figure 2: (a) Comparison between energies of single particle electron-hole pair (no Coulomb interaction) and exciton. (b) Comparison between exciton (Ex) and bi-exciton (Exx) energies calculated with and without the polarization effect. (c) Comparison between energies of bi-excitons using full CI calculations and the Hartree approximation equivalent.

4. Summary and conclusions

In conclusion we calculated exciton and bi-exciton energies in type-I CdSe and type-II CeSe/CdTe colloidal quantum dots. We showed that the electron-hole pairs indeed interact strongly and it is necessary to take into account the Coulomb interaction in energy level calculations. We proved that inclusion of the polarization effect is important because of the large dielectric constant mismatch at the dot/environment boundary. Finally, we showed that the configuration interaction is significant in the calculations of bi-exciton energies. Our theoretical results of ground state exciton energy agree very well with experimental measured results for CdSe QDs of radius 1.5 nm [11].

5. Acknowledgements

This research is funded by: EPSRC grant "Enhanced multiple exciton generation in colloidal quantum dots" and the Royal Society grant "High Performance Computing in Modeling of Innovative Photo-Voltaic Devices". We acknowledge help from the N8 Research Partnership and Science and Technology Facilities Council for providing the computational resources used to conduct this research. We would also like to express our gratitude to Prof Wendy Flavell, Prof Paul O'Brien and Thomas Walsh for useful discussions.

References

[1] Califano M, Zunger A and Franceschetti A 2004 *Applied Physics Letters* **84** 2409–2411 ISSN 0003-6951, 1077-3118
[2] Tomić S 2010 *Physical Review B* **82** 195321
[3] Wang L W, Califano M, Zunger A and Franceschetti A 2003 *Physical Review Letters* **91** 056404
[4] Tomić S, Sunderland A G and Bush I J 2006 *Journal of Materials Chemistry* **16** 1963–1972 ISSN 1364-5501
[5] Vukmirović N, Ikonić Z, Indjin D and Harrison P 2006 *Journal of Physics: Condensed Matter* **18** 6249 ISSN 0953-8984
[6] Vukmirovć N and Tomić S 2008 *Journal of Applied Physics* **103** 103718–103718–12 ISSN 00218979
[7] Tomić S, Montanari B and Harrison N M 2008 *Physica E: Low-dimensional Systems and Nanostructures* **40** 2125–2127 ISSN 1386-9477
[8] Piryatinski A, Ivanov S A, Tretiak S and Klimov V I 2007 *Nano Letters* **7** 108–115 ISSN 1530-6984
[9] Franceschetti A, Williamson A and Zunger A 2000 *The Journal of Physical Chemistry B* **104** 3398–3401 ISSN 1520-6106
[10] Bolcatto P G and Proetto C R 2001 *Journal of Physics: Condensed Matter* **13** 319 ISSN 0953-8984
[11] Yu W W, Qu L, Guo W and Peng X 2003 *Chemistry of Materials* **15** 2854–2860

THz intervalence band antipolaritons

Inuwa A Faragai and Mauro F Pereira

Materials and Engineering Research Institute, Sheffield Hallam University, S1 1WB, Sheffield, UK

E-mail: m.pereira@shu.ac.uk

Abstract. THz polaritons and antipolaritons have strong potential for device applications and are a challenging field of fundamental studies. In this paper, we start from a numerically exact nonequilibrium many body solutions and adjust it to a simplified nonlinear dielectric constant approach to the optical susceptibility. The resulting expression is inserted in the wave equation to describe the coupling of TE-polarized THz radiation with an intervalence band transition in GaAs/$Al_{0.3}Ga_{0.7}As$ multiple quantum wells embedded in microcavities. The energy dispersions relations leading to THz polaritons are investigated. Here we focus on the impact of dephasing and scattering processes for different structures and excitation conditions in an inverted medium leading to antipolaritons.

1. Introduction

Intersubband based THz polariton emitters are attracting major research interest [1,2]. The coupling of THz radiation with intersubband transitions in microcavities can lead to further tunability and improved quantum efficiency and may play a major role in future THz materials and devices [3]. In this paper we extend and complement the results of Refs. [4,5] by investigating the influence of dephasing and scattering mechanisms on TE mode, THz antipolaritons in different structures based on GaAs/$Al_{0.3}Ga_{0.7}As$ multiple quantum wells. As in the case of conduction band transitions, polaritons [6-8] and antipolaritons [9] can be realized with either absorptive (passive) or gain (inverted) media. However, in contrast to interconduction band transitions, both transverse electric (TE) and transverse magnetic (TM) polarized cavity modes [10] are possible.

2. Mathematical Approach and Model Equations

The input to the dielectric constant used here is obtained by self-consistent evaluation of manybody Nonequilibrium Green's Functions (NGF). The numerical method is summarized as follows. The THz susceptibility $\chi(\omega)$ is extracted from Keldysh Green's Functions,

$$\left[\hbar\omega - e_{\mu\nu}(k) + i\Gamma_{\mu\nu}\right]\chi_{\nu\mu}(k,\omega) - \Delta n_{\nu\mu}(k)\sum_{\vec{q}\neq\vec{k}}\chi_{\nu\mu}(k',\omega)V^{\nu\mu}_{\vec{q}-\vec{k}} = \wp_{\nu\mu}(k)\Delta n_{\nu\mu}(k). \quad (1)$$

Here $\wp_{\nu\mu}(k)$and $\Delta n_{\nu\mu}(k)$ are the transition dipole moment and the nonequilibrium population difference between hole subbands ν and μ. More details of the renormalized energies $e_{\mu\nu}(k)$, electron-electron scattering $\Gamma_{\mu\nu}$ and Coulomb matrix elements $V^{\nu\mu}_{\vec{q}-\vec{k}}$ are given in Refs. [11-17]. The system is globally out of equilibrium. The holes are thermalized in the upper hole subband (this is the electron-hole and not the conduction-valence band picture) and the lowest hole band is considered to be empty. The dielectric function is obtained by adjusting the full numerical solution to a Lorentzian, namely

$$\varepsilon(\omega) = \varepsilon_b + 4\pi\lambda\chi(\omega) = 1 - \frac{\lambda}{\varepsilon_b}\sum_{\mu\nu}\frac{\Lambda_{\mu\nu}}{\omega-\omega_{\mu\nu}+i\gamma_{\mu\nu}} - \frac{\Lambda_{\mu\nu}}{\omega+\omega_{\mu\nu}+i\gamma_{\mu\nu}} \quad (2)$$

$\chi(\omega)$ is the total optical susceptibility, ε_b, ω, and $\omega_{\mu\nu}$ are background dielectric constant, photons and transition frequencies between subbands $\mu \neq \nu$ respectively. The constant λ is given by $\lambda = N_w \frac{L_c}{L_w}$ [14], where N_w is the number of QWs, L_w and L_c denote respectively the QW and cavity core lengths. Here $\gamma_{\mu\nu}$ is the dephasing term, while $\Lambda_{\mu\nu}$ is due to dipole coupling and measures the oscillator strength of the transition. The dispersion relations shown in Fig.2 are obtaining by substituting Eq. (2) into Maxwells' equations [9,10]. Introducing $y = \frac{\omega}{\omega_o}$, $\Omega_c = \frac{\omega_c}{\omega_0} = \frac{c\pi}{\omega_o L_c}\left(\frac{1}{\varepsilon_b}\right)^{\frac{1}{2}}$ for the cavity resonance and ω_o , the reference resonance frequency (for one transition between subbands 1 and 2, $\omega_0 = \omega_{12}$ ε_r and ε_i , we obtain the dispersion relation

$$\sin\theta_i = \left(\frac{\varepsilon_b}{\varepsilon_s}\right)^{\frac{1}{2}}\sqrt{1-\lambda\Phi-\frac{\Omega_c^2}{y^2}}\ ,\ \Phi = \sum_{\mu\nu}\frac{\Delta_{\mu\nu}(y-\Omega_{\mu\nu})}{(y-\Omega_{\mu\nu})^2+\Gamma^2} - \frac{\Delta_{\mu\nu}(y+\Omega_c)}{(y+\Omega_c)^2+\Gamma^2}\ ,\ \Delta_{\mu\nu} = \frac{\Lambda_{\mu\nu}}{\varepsilon_b\omega_o} \text{ and } \Gamma = \frac{\gamma_{\mu\nu}}{\omega_o}. \quad (3)$$

The geometry considered in this work is shown in Fig. 1. In order to ensure that incident light reaches the sample in the cavity with a suitable incidence angle, a prism made of a high refractive material is placed on top of the sample.

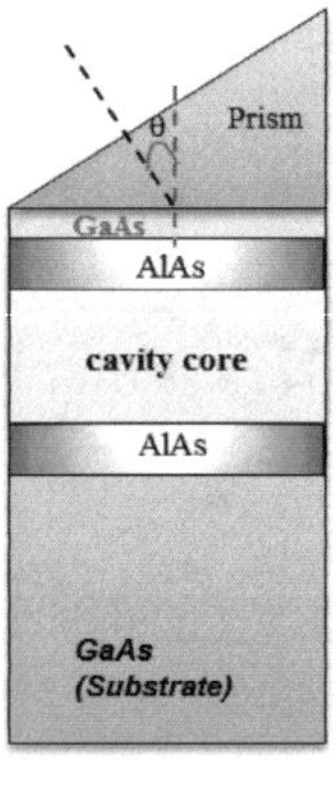

Figure 1. Sample structure. Light is confined through total internal reflection by low refractive index AlAs layers below and above the cavity core, which contains 170 GaAs quantum wells, each separated by 150nm $Al_{0.3}Ga_{0.7}As$ barrier except the last top barrier layer which was set to 250nm and these lead to cavity lengths L_C equal to 26.6μm and 28.5μm for the 5nm and 10nm MQW structures respectively. The top and bottom layers of AlAs are both set to 1300 nm and then completed the sample with a top 5nm GaAs to avoid AlAs oxidation due to air interface. The corresponding cavity parameters are: cavity resonances ω_c = 7.3579meV / 6.8568meV; background dielectric constant ε_b = 10.0117 / 10.0392 and the effective medium constant λ = 0.0319 / 0.0596 for 5nm / 10nm MQW samples.

2. Numerical Results and Discussion

THz antipolaritons due to interacting TE- polarized cavity modes and intervalence band transitions in the gain regime are shown next for two different samples, in which the cavities have either 10nm or 5nm $GaAs/Al_{0.3}Ga_{0.7}As$ MQWs. Figure 2 shows the influence of dephasing 'γ' and the effect of changing the carrier density 'ΔN', on the THz intervalence band antipolariton dispersions for a sample following the geometric structure shown in Fig. 1 for 5nm ((a) and (b)) and 10nm ((c) and (d)) MQWs respectively. Since the lowest band is assumed to be empty, this also corresponds to the population difference. $\mu, \nu = 1,2$ denote respectively the first and second subbands. $\Lambda_{\mu\nu}$ and $\omega_{\mu\nu}$ are obtained by adjusting the full numerical manybody expression to the Lorentzian approximation in Eq. 2. The increasing dephading for fixed carrier densities simulates time resolved experiments in which the onset of dephasing can be monitored. The dephasing increases from the outer curves towards the centre. Analogously to the interband exciton [18] and the TM mode inter-conduction band cases [19],

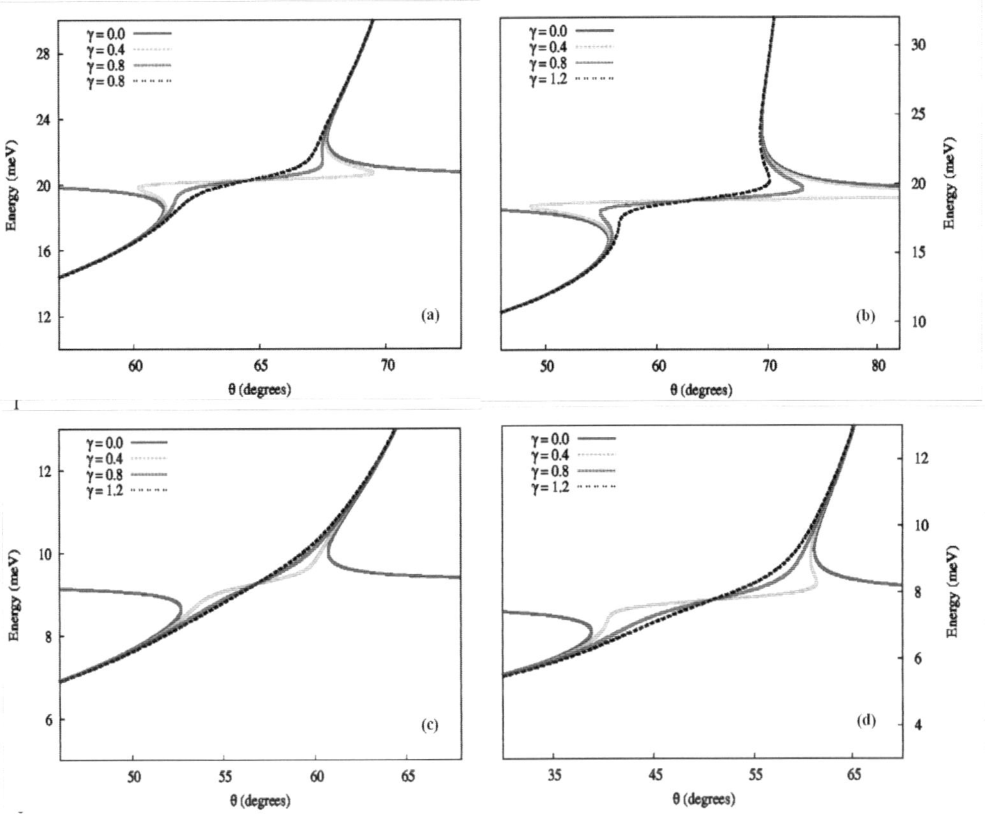

Figure 2. THz intervalence band antipolaritons due to coupling of a TE polarized cavity mode and intervalence band excitations: (a) Lw = 5nm, $\Delta N = 1.0\times10^{11}$ cm^{-2} $\Lambda_{\mu\nu} = -8.4634$ meV, $\omega_{\mu\nu} = 20.2753$. (b) Lw = 5nm, $\Delta N = 4.0 \times 10^{11}$ cm^{-2}, $\Lambda_{\mu\nu} = -32.7163$ meV, $\omega_{\mu\nu} =18.7734$ me. (c) Lw = 10nm, $\Delta N = 1.0\times10^{11}$ cm^{-2}, $\Lambda_{\mu\nu} = -25.5945$, $\omega_{\mu\nu} = 7.7596$. (d) Lw = 10nm, $\Delta N = 4.0 \times 10^{11}$ cm^{-2}, $\Lambda_{\mu\nu} = -25.5945$, $\omega_{\mu\nu} = 7.7596$.

Note that the 5 nm MQWs have transitions with larger transition energies than the 10 nm MQWs so for the same dephasing, the antipolariton features are stronger. Compare the dashed lines in both cases. For larger densities, the absolute value of the longitudinal-transverse (LT) splitting $\Delta_{\mu\nu}$ increases. This increased oscillator strength leads to a stronger separation of the two branches. Note however the interplay of larger quantum well width (and thus larger λ factor) and larger oscillator strengths that make the branch separation comparable for the same density. In other words, even though $\Delta_{\mu\nu}$ is larger for the 5 nm MQWs, due to better electronic overlap, λ is larger in the structure containing 10 nm MQWs, roughly equilibrating the effect. This study further suggests that samples with different well widths in the active region might lead to larger tunability (see the different energy ranges) in applications where the antipolariton coupling might be used.

In we have investigated the interaction of TE Mode THz radiation and intervalence band transitions in quantum well gain media. This can lead to interesting THz based devices based on TE modes in contrast with inter-conduction band transitions that only couple to TM polarized radiation. A Nonequilibrium Many Body Approach for the optical response beyond the Hartree-Fock approximation is used as input to the effective dielectric function formalism for the polariton problem. The energy dispersions relations in the THz range are obtained by adjusting the full numerical solutions to simple analytical expressions to improve clarity, allowing a broad audience to investigate similar phenomena. The method allows to combine predictive microscopic optical responses with a simple dielectric model for the cavity leading to a powerful design tool for THz polaritonics.

Acknowledgements
The authors acknowledge support from COST ACTIONs MP1204 and BM1205. I.A. Faragai's research is supported by the KUST-Wudil under TETFUND, Nigeria

References

[1] S. De Liberato, C. Ciuti and C. C. Phillips, Terahertz lasing from intersubband polariton-polariton scattering in asymmetric quantum wells, Phy. Rev. B. 87, 241304(R) (2013).

[2] M. Geiser, G. Scalari, F. Castellano, M. Beck and J. Faist, Room temperature terahertz polariton emitter Applied Physics Letters 101, 141118 (2012).

[3] M.F. Pereira, TERA-MIR radiation: materials, generation, detection and applications, Opt Quant Electron 46, pp. 491–493 (2014).

[4] I. A. Faragai and M. F. Pereira, Interaction of valence band excitations and terahertz TE-polarized cavity modes, Optical and Quantum Electronics, Opt Quant Electron 46, pp. 527–531, (2014).

[5] M.F. Pereira and I.A. Faragai, Coupling of THz with intervalence band transitions in microcavities, Optics Express, Vol. 22, Issue 3, pp. 3439-3446 (2014).

[6] D. Dini, R. Köhler, A. Tredicucci, G. Biasiol, L. Sorba, and F. Beltram, "Microcavity Polariton Splitting of Intersubband Transitions", Phys. Rev. Lett. 90, 116401 (2003).

[7] A. A. Anappara, A. Tredicucci and F. Beltram, Cavity polaritons from excited-subband transitions, App. Phys. Lett., 91, 231118 (2007).

[8] S. de Liberato,and C. Ciutti, Stimulated scattering and lasing of intersubband cavity polaritons, Phys. Rev. Lett. 102, 136403 (2009)

[9] M. F. Pereira Jr. Intersubband antipolaritons: Microscopic approach, Phys. Rev. B 75, 195301 (2007).

[10] M.F. Pereira Jr. and H. Wenzel, Interplay of Coulomb and nonparabolicity effects in the intersubband absorption of electrons and holes in quantum wells, Phys. Rev. B70, 205331 (2004

[11] M. F. Pereira Jr., Intervalence transverse-electric mode terahertz lasing without population inversion, Phys. Rev. B 78, 245305 (2008).

[12] T. Schmielau and M.F. Pereira Jr., Nonequilibrium many body theory for quantum transport in terahertz quantum cascade lasers, Appl. Phys. Lett. 95, 231111 (2009).

[13] T. Schmielau and M.F. Pereira, Impact of momentum dependent matrix elements on scattering effects in quantum cascade lasers, physica status solidi b 246, 329 (2009).

[14] M.F. Pereira Jr, R. Nelander, A. Wacker, D.G. Revin, M.R.Soulby, L.R. Wilson, J.W. Cockburn, A.B. Krysa, J.S. Roberts, and R.J. Airey, Characterization of Intersubband Devices Combining a Nonequilibrium Many Body Theory with Transmission Spectroscopy Experiments, Journal of Materials Science: Materials in Electronics 18, 689 (2007).

[15] M.F. Pereira Jr, R. Nelander, A. Wacker, D.G. Revin, M.R.Soulby, L.R. Wilson, J.W. Cockburn, A.B. Krysa, J.S. Roberts, and R.J. Airey, "Characterization of Intersubband Devices Combining a Nonequilibrium Many Body Theory with Transmission Spectroscopy Experiments", Journal of Materials Science: Materials in Electronics 18, 689 (2007).

[16] M. F. Pereira and S. Tomić, Intersubband gain without global inversion through dilute nitride band engineering, Appl. Phys. Lett. 98, 061101 (2011)

[17] M.F. Pereira Jr., Microscopic approach for intersubband-based thermophotovoltaic structures in the THz and Mid Infrared, JOSA B Vol. 28, Iss. 8, pp. 2014–2017 (2011).

[18] A. Quattropani, W. Czaja, Physica Scripta T29, 162 (1989)

[19] M.F. Pereira, The influence of dephasing in the coupling of light with intersubband transitions, Microelectronics Journal 40, 841 (2009).

Impact of different electron-phonon scattering models on the electron transport in a quantum wire

A Price[1], A Martinez[1], R Valin[1] and J R Barker[2]
[1] College of Engineering, Swansea University, UK;
[2] School of Engineering, University of Glasgow, UK.

E-mail: A.C.PRICE.625036@swansea.ac.uk

Abstract. The non-equilibrium Green's function (NEGF) approach has been widely used to describe quantum transport in nanostructures. In general, the Electron-phonon (e-ph) interaction is regarded to be local in space, as transport simulations using non-local e-ph interactions are very computationally intensive. As such non-local interactions are rarely used to simulate nanotransistors. In this work, non-local e-ph interactions in the self-consistent Born approximation (SCBA) are used to simulate transport through a quantum wire. The non-local model is compared with the results for local, nearest-neighbour and second nearest-neighbour interactions. It was found that for a $40nm$ quantum wire with a dephasing factor of $D_0 = 0.001eV^2$ and an applied potential of $0.1V$, the local model underestimates the current by 20%. Furthermore, to study electron transport in Graphene and polymers, non-local e-ph interactions must be considered.

1. Introduction

Non-equilibrium Green's functions (NEGF) provide a powerful formalism when studying electron transport in nanodevices. The Electron-phonon interaction is usually approximated as being local in space [1, 2]. For many devices this approximation is not valid. In particular, when considering polar phonons in crystals a non-local model needs to be implemented [3]. Four different dephasing models have been investigated for scattering in a one-dimensional quantum wire: local, nearest-neighbour, second nearest-neighbour and non-local. A schematic of the device and the range of the four models is given in figure 1.

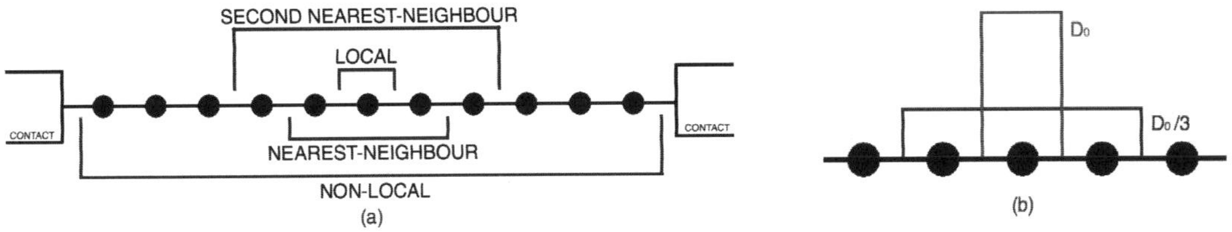

Figure 1. (a) Schematic of the device showing the range for local, nearest-neighbour, second nearest-neighbour and non-local interactions. (b) Illustration of the scaling of the dephasing factor D_0 for local and nearest-neighbour interactions.

Journal of Physics: Conference Series **526** (2014) 012007 doi:10.1088/1742-6596/526/1/012007

2. Theory

The system is described by the electronic Green's functions G and self-energies Σ, which are computed using the Born iteration. The self-energies are initially set to zero. The retarded Green's function is calculated using Dyson's equation

$$G^r(\omega) = [\omega 1 - H - \Sigma^r(\omega)]^{-1} \tag{1}$$

and the lesser-than and greater-than Green's functions using the quantum kinetic equations

$$G^{\lessgtr}(\omega) = G^r(\omega)\Sigma^{\lessgtr}(\omega)G^a(\omega) \tag{2}$$

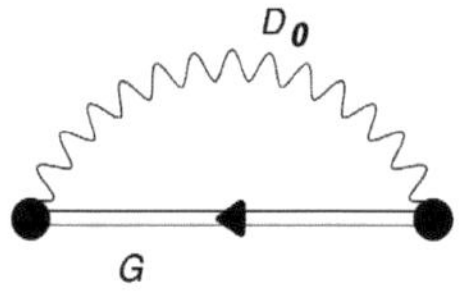

Figure 2. Fock diagram for the electron-phonon interaction in SCBA

The total self-energies are given by the sum of the contribution of the two contacts and the self-energy due to scattering

$$\Sigma^x(\omega) = \Sigma_1^x(\omega) + \Sigma_2^x(\omega) + \Sigma_S^x(\omega) \tag{3}$$

where $x = \{r, a, <, >\}$. The new values for the self-energies are used to re-calculate the Green's functions and this process is repeated until the solution converges. Once the solution has converged observables such as the current in the contacts can be calculated. Our choice for the scattering self-energy is given by the self-consistent Born approximation (SCBA) [4]

$$\Sigma_S^{\lessgtr}(\omega) = id_0 \int_{-\infty}^{\infty} \frac{d\omega'}{2\pi} D_0^{\lessgtr}(\omega - \omega')G^{\lessgtr}(\omega') \tag{4}$$

$$\Sigma_S^r(\omega) = id_0 \int_{-\infty}^{\infty} \frac{d\omega'}{2\pi} [D_0^r(\omega - \omega')G^r(\omega') + D_0^r(\omega - \omega')G^<(\omega') + D_0^<(\omega - \omega')G^r(\omega')] \tag{5}$$

The SCBA satisfies current conservation. All results henceforth are current conserving.

3. Results

Figure 3 (a) compares the current in a quantum wire for the four dephasing models and ballistic transport for wires of different channel length. For the non-local model the dependence of the current on the length of the device is constant. This is because non-local scattering introduces momentum conserving dephasing, which relaxes the phase of the electrons while conserving the momentum [5]. For the local model the current decreases rapidly with increasing length. Local interactions introduce momentum relaxing dephasing, which relaxes the phase and the momentum of the electrons, such that momentum relaxation is equal to the energy relaxation. Nearest-neighbour and second nearest-neighbour interactions exhibit a combination of momentum conserving and momentum relaxing dephasing. Figure 3 (b) shows how the current varies for a $27nm$ wire for different applied fields. Crucially, figure 3 (b) demonstrates that the local model underestimates the current. For applied fields up to $\vec{E} = 3 \times 10^4 V/cm$ the current from the local model is 16% less than the current produced by non-local dephasing. This difference increases with length to approximately 20% for a $40nm$ chain. A field of

Journal of Physics: Conference Series **526** (2014) 012007 doi:10.1088/1742-6596/526/1/012007

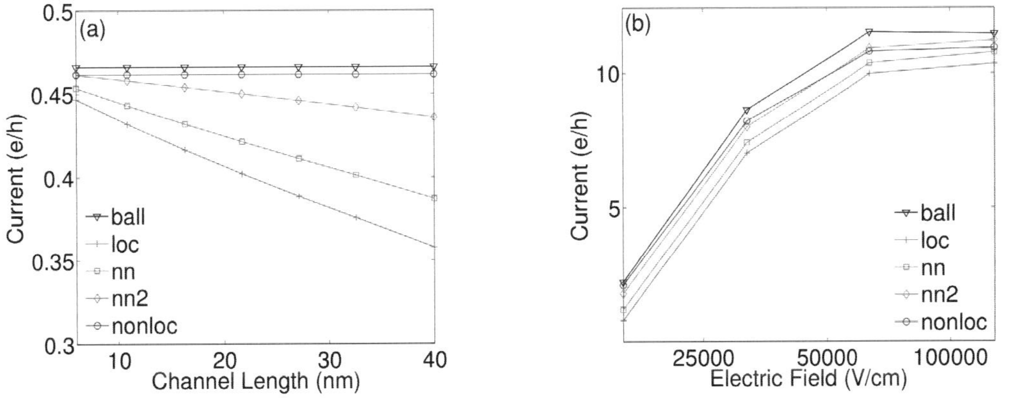

Figurc 3. (a) Comparison of the current in a quantum wire for ballistic (ball), local (loc), nearest-neighbour (nn), second nearest-neighbour (nn2) and non-local (nonloc) transport for wires of different channel length. Dephasing factor $D_0 = 0.001eV^2$. Applied electric field $\vec{E} = 2 \times 10^3 V/cm$. (b) Current vs. Electric field for a $27nm$ wire. Dephasing factor $D_0 = 0.001eV^2$.

$\vec{E} = 3 \times 10^4 V/cm$ corresponds to an applied potential of $0.1V$ for a $40nm$ device, which is typical for a nanotransistor. For fields greater than this the non-local current begins to decrease in relation to the other three dephasing models, with the non-local current becoming less than the second nearest-neighbour current at $\vec{E} > 5 \times 10^4 V/cm$. This effect can be understood by considering the local density of states (LDOS) and energy-resolved current for low and high electric field.

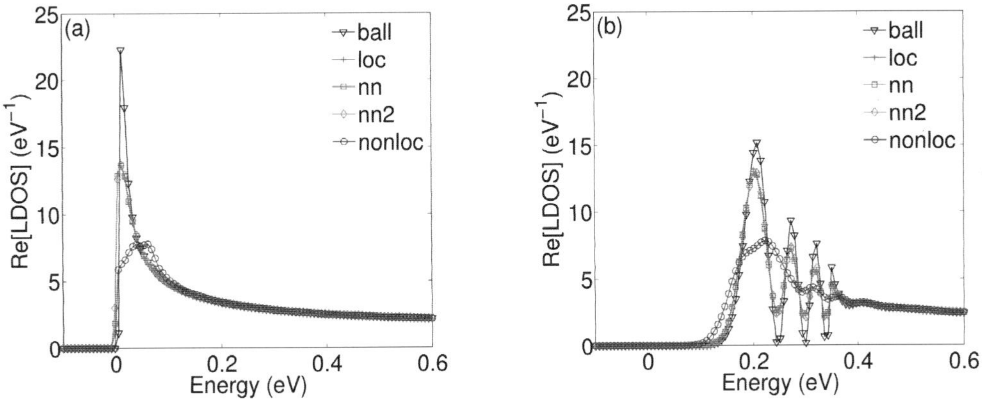

Figure 4. (a) Local density of states in the middle of the channel for $27nm$ device for the four dephasing models and ballistic transport, with an applied electric field of (a) $\vec{E} = 8 \times 10^3 V/cm$. (b) $\vec{E} = 1.28 \times 10^5 V/cm$. Dephasing factor $D_0 = 0.001eV^2$.

Figure 4 compares the local density of states (LDOS) for the four dephasing models and ballistic transport. The LDOS is the same for local, nearest-neighbour and second nearest-neighbour models due to the sum rule. The LDOS for non-local model is broadened because the non-local model has a greater amount of dephasing than the local model for the same D_0, but the total number of states is the same for each model. For the high electric field in figure

4 (b), there are oscillations in the LDOS in the channel for all models. This is a result of the electrons travelling from the drain being reflected by the potential profile which forms a barrier. This effect is more prevalent at higher electric fields due to the potential barrier being larger; creating more reflections.

Figure 5 compares the energy resolved current for low and high electric field. In figure 5 (a) for low electric field, the current for the four dephasing models and ballistic transport flows at the same energy. However, in figure 5 (b) for high electric field, the non-local current maximum is shifted to an energy $14meV$ greater than the second nearest-neighbour current maximum. This in turn results in a reduction in the current due to there being less occupied states at higher energies.

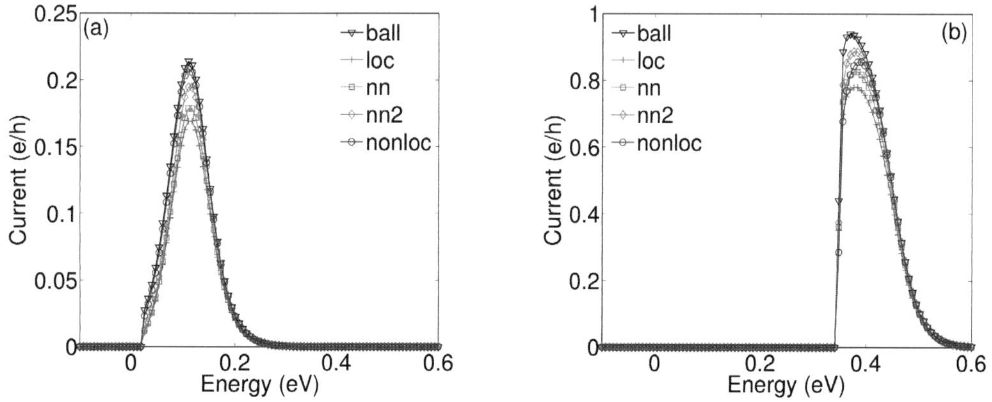

Figure 5. Current in the source as a function of energy for a $27nm$ device for the four dephasing models and ballistic transport, with an applied electric field of (a) $\vec{E} = 8 \times 10^3 V/cm$. (b) $\vec{E} = 1.28 \times 10^5 V/cm$. Dephasing factor $D_0 = 0.001eV^2$.

4. Conclusions

The NEGF formalism has been applied to a 1D system, comparing local, nearest-neighbour, second nearest-neighbour and non-local interactions. Non-local and local interactions have been seen to introduce momentum conserving and momentum relaxing dephasing respectively. While nearest-neighbour and second nearest-neighbour interactions exhibit a combination of momentum conserving and momentum relaxing dephasing. For devices $< 20nm$ with an applied field $\vec{E} < 3 \times 10^4 V/cm$, the second nearest-neighbour model provides a good approximation for the computationally intensive non-local model. For $\vec{E} > 5 \times 10^4 V/cm$ the current for the second nearest-neighbour model is greater than the non-local current. This is due to the current for the non-local model flowing at a higher energy; reducing the number of available states that can contribute to the transport. Non-local interactions heavily contribute to the scattering rate of polar phonons, and also affect inelastic scattering of acoustic phonons in short devices [3].

References

[1] Svizhenko A and Anantram M P 2005 *Phys. Rev.* B **72** 085430
[2] Jin S, Park Y J and Min H S 2006 *J. Appl. Phys.* **99** 123719
[3] Kubis T C 2009 *Quantum transport in semiconductor nanostructures* (PhD Thesis)
[4] Lü J T and Wang J S 2007 *Phys. Rev.* B **76** 165418
[5] Golizadeh-Mojarad R and Datta S 2007 *Phys. Rev.* B **75** 081301

Journal of Physics: Conference Series **526** (2014) 012008 doi:10.1088/1742-6596/526/1/012008

Generalised effective mass theory of sub-surface scanning tunnelling microscopy: application to weakly bound impurity states

Mervyn Roy and P A Maksym

Department of Physics and Astronomy, University of Leicester, Leicester, LE1 7RH, UK

E-mail: mr6@le.ac.uk

Abstract. We apply our generalised effective mass theory of sub-surface scanning tunnelling microscopy (STM) (Phys. Rev. B **19**, 195304 (2010)) to simulate STM images of electronic states localised around sub-surface Si dopant atoms in GaAs. In the case of these shallow impurity-states, we demonstrate that electrostatic effects from image-charges and from the STM tip have a strong influence on the sub-surface state and hence the simulated image.

1. Introduction

Scanning tunnelling microscopy (STM) samples the local density of states (LDOS) above a surface, but this is influenced by sub-surface structure, and there is growing interest in using STM to probe quantum states localised below a surface. States associated with near-surface shallow donors and cleaved dots have all been recently imaged [1, 2, 3].

To understand these experiments we must calculate the LDOS, but such calculations are challenging because the sub-surface states can be very broad, extending over millions of atoms. In such large systems an effective mass approach is appealing as it provides an excellent semi-quantitative description of shallow-impurity states in the bulk. However, because it is a bulk theory, a traditional effective mass approach cannot be used to find the LDOS above a surface.

Here we discuss a recent extension to the theory and its application to STM. In Ref. [1] we demonstrated that the LDOS above the surface can be written in terms of a simple weighted integral of the bulk envelope function. The resulting 'surface envelope function' contains the lateral information in the tail of the LDOS that is probed in STM. We apply this theory to simulate STM images of electronic states localised around sub-surface Si dopant atoms in GaAs.

We develop a detailed continuum model of the sub-surface impurity, including the band bending caused by the STM tip, and the effect of the image-charges. These electrostatic effects are important for the shallow impurity-states and contribute to the observed increase in binding energy with decreasing dopant depth [2].

2. Generalised effective mass theory

In a recent paper [1] we generalised effective mass theory to find the local density of states above a semiconductor surface in the presence of a sub-surface nanostructure or impurity. Here we quote the key results appropriate to GaAs (110): we give the single-band results for a crystal with an inversion plane parallel to the surface. Details and general results are given in Ref. [1].

Journal of Physics: Conference Series **526** (2014) 012008 doi:10.1088/1742-6596/526/1/012008

In our model, a slab of semiconductor, of lattice constant a_0, extends from $-L$ to 0 in the z-direction and is of infinite extent parallel to the surface. We consider a slowly varying confinement potential, $v(\boldsymbol{\rho},z)$, that is generated by an impurity or nanostructure located below the surface, close to $z = 0$. The localised states, $\psi(\boldsymbol{\rho}, z)$, associated with $v(\boldsymbol{\rho},z)$ then have some tail above the surface that can be probed in STM. The experimental tunnelling current is proportional to $|\psi(\boldsymbol{\rho}, z)|^2$ at $(\boldsymbol{\rho}_t, z_t)$, the position of the tip [4]. So, to compare to STM experiments, we simply need to calculate the localised states where $z > 0$.

The single-band wavefunction above the surface can be written [1],

$$\psi(\boldsymbol{\rho}, z) = F_s(\boldsymbol{\rho}) w_{00}(\boldsymbol{\rho}, z), \tag{1}$$

where $F_s(\boldsymbol{\rho})$ is a surface envelope function that varies in the plane parallel to the surface and $w_{00}(\boldsymbol{\rho}, z)$ is a Bloch-like function at the band edge, with wave vector $\mathbf{k}_{||} = 0$, and $k_z = 0$. As in the traditional bulk effective mass theory, the Bloch part of ψ describes the atomic scale variation that comes from the host semiconductor in the absence of $v(\boldsymbol{\rho},z)$. The surface envelope function then contains all the effects of the sub-surface nanostructure or impurity. It is given by

$$F_s(\boldsymbol{\rho}) = \int_{-L}^{0} G(z) F(\boldsymbol{\rho}, z) dz, \tag{2}$$

where $F(\boldsymbol{\rho}, z)$ is a bulk envelope function that can be found by diagonalising the usual bulk effective mass Hamiltonian with the additional boundary conditions that $F(\boldsymbol{\rho}, z) = 0$ at the edges of the slab where $z = -L$ and $z = 0$. In the region $|z| \ll L$, where F is large, the weighting function $G(z)$ is simply

$$G(z) = \frac{2}{z} \sin^2\left(\frac{\pi z}{2a_0}\right). \tag{3}$$

In an STM experiment, the tunnelling current is $I_{STM} \propto |F_s(\boldsymbol{\rho}_t)|^2 |w_{00}|^2 = I(\boldsymbol{\rho}_t)|w_{00}|^2$. So, to calculate the atomic scale variation in tunnelling current we must find w_{00} [5]. However, for qualitative or semi-quantitative analysis of the states associated with the sub-surface impurity or nanostructure it is often sufficient to calculate only the large scale variation in I_{STM}. In this case the procedure is very simple:

(i) calculate the bulk envelope function, $F(\boldsymbol{\rho}, z)$, with the additional boundary conditions that $F = 0$ at the surfaces of the slab.

(ii) calculate the surface envelope function, $F_s(\boldsymbol{\rho})$, from Eq. (2).

(iii) calculate the tunnelling current $I(\boldsymbol{\rho}_t) \propto |F_s(\boldsymbol{\rho}_t)|^2$.

3. Effective mass model of a sub-surface Si in GaAs

We consider a Si dopant located at $z = -z_d$ in a semi-infinite GaAs slab that extends from $z = -\infty$ to $z = 0$. To find the bulk envelope function (step (i) from section 2) we solve the Schrödinger equation in cylindrical polar co-ordinates,

$$\left(\frac{-\hbar^2}{2m^*}\left[\frac{1}{\rho}\frac{\partial}{\partial \rho}\left(\rho\frac{\partial}{\partial \rho}\right) + \frac{\partial^2}{\partial z^2}\right] + v(\rho, z)\right) F(\rho, z) = EF(\rho, z), \tag{4}$$

where $F(\rho, z) = 0$ at $z = 0$ and $F \to 0$ as $z \to -\infty$. Here we only consider systems with cylindrical symmetry and, in Eq. (4), v is cylindrically symmetric in the absence of an STM tip and in the presence of a cylindrically symmetric tip centered at $\boldsymbol{\rho} = 0$.

In the bulk, when $z_d \ll 0$, the potential associated with a Si dopant is simply the usual Coulomb potential screened by the macroscopic dielectric constant of GaAs,

$v_{imp}(\rho, z) = -e^2/(4\pi\epsilon_0\epsilon_r)/\sqrt{\rho^2 + (z - z_d)^2}$, where $\epsilon_r = 13.1$. When z_d is close zero, however, the surface affects the electric field and we include this effect using the classical method of images. The potential may also be affected by the STM tip, so that

$$v(\rho, z) = v_{imp}(\rho, z) + v_{di}(\rho, z) + v_{ei}(\rho, z) + v_{bb}(\rho, z), \tag{5}$$

where $v_{di} = -e^2(\epsilon_r - 1)/(4\pi\epsilon_0\epsilon_r(\epsilon_r + 1))/\sqrt{\rho^2 + (z + z_d)^2}$ is the dopant image potential, and $v_{ei} = e^2(\epsilon_r - 1)/(16\pi\epsilon_0\epsilon_r(\epsilon_r + 1))/|z|$ is the electron image potential [7]. In the presence of an STM tip centred at $\boldsymbol{\rho}_t = 0$, $v_{bb}(\rho, z)$ gives the effect of the tip induced band bending (TIBB). We calculate the TIBB by solving Poisson's equation using the standard finite-difference code of Feenstra [6]. We simulate an experimentally realistic [2, 3] ultrasharp tip of opening angle 19°, and tip radius 2.5 nm, with a tip voltage of $V_{tip} = -0.5$ V. The tip is 0.5 nm from the GaAs surface and the GaAs has an average Si doping concentration of 10^{18} cm^{-3}.

We solve Eq. (4) by expanding $F(\rho, z)$ in a basis of harmonic oscillator functions and then diagonalising the resultant Hamiltonian matrix [8].

$$F(\rho, z) = \sum_{n}^{n_{max}} \sum_{m}^{m_{max}} a_{nm} Z_{2m+1}(z) R_{n0}(\rho), \tag{6}$$

where $Z_m(z) = \exp(-z^2/2z_0^2)H_m(z/z_0)/(2^m m!\sqrt{\pi}z_0)^{1/2}$ are the 1D oscillator functions, $R_{n0}(\rho) = \exp(\rho^2/4\rho_0^2)L_n(\rho^2/2\rho_0^2)/\rho_0$ are 2D oscillator functions with zero angular momentum, and L_n and H_m are Laguerre and Hermite polynomials respectively. In the expansion in Eq. (6) we only include the 1D oscillator states with odd quantum number (and hence odd symmetry about $z = 0$) to ensure that $F(\rho, z)$ goes to zero at the surface.

4. Results and discussion

Here we focus on a single silicon dopant at $z_d = -0.6$ nm. To investigate the effect of the image charges and the TIBB on the localised state we calculate $F(\rho, z)$ for 3 different cases: in (i) we neglect the image potentials and set $v_{bb} = 0$; in (ii) we include the image potentials and set $v_{bb} = 0$; in (iii) we include both image potentials and $v_{bb}(\rho, z)$. In all 3 cases we converge the ground state energy to within 0.05 meV with 96 states in our expansion ($n_{max} = 12$, $m_{max} = 16$), using oscillator length scales of $z_0 = 19$ nm, and $\rho_0 = 11$ nm.

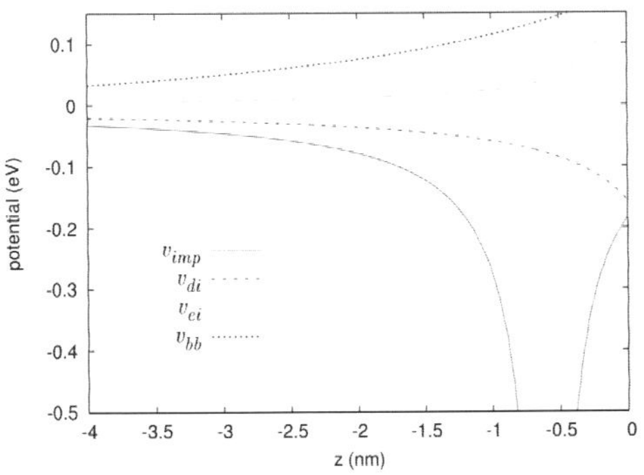

Figure 1. Components of $v(\rho, z)$ shown at $\rho = 0$ as a function of z. The effect of the dopant image potential, v_{di}, is to deepen the confinement generated by v_{imp}, while the electron image potential, v_{ei}, repels the state away from the surface. The TIBB (v_{bb}) ($V_{tip} = -0.5$ V) raises the total potential and acts to repel the state away from the tip at $\rho_t = 0, z_t = 0.5$ nm.

Figure 1 shows line scans through $v_{imp}(\rho, z), v_{di}(\rho, z), v_{ei}(\rho, z)$ and $v_{bb}(\rho, z)$. Figure 2 shows $|F(\rho, z)|^2$ as a function of ρ and z for the 3 different cases. In case (i) (left panel) we find a binding energy of $E_b = |E| = 1.4$ meV; the presence of the surface reduces the binding energy below the well-known bulk value of 5.3 meV. In the centre panel of Fig. 2 we see that the result of including the image potentials (case (ii)) is to localise the state closer to $z = 0$, $\rho = 0$, increasing the binding energy from 1.4 to 3.2 meV. However, although the effect on E_b is significant, it is not large enough to explain the experiment [2] where E_b was estimated to increase to $\sim$ 10 meV, around twice the bulk value. The right panel shows $|F(\rho, z)|^2$ in case (iii). The TIBB pushes the state away from the surface, decreasing the binding energy to $E_b = 2.4$ meV.

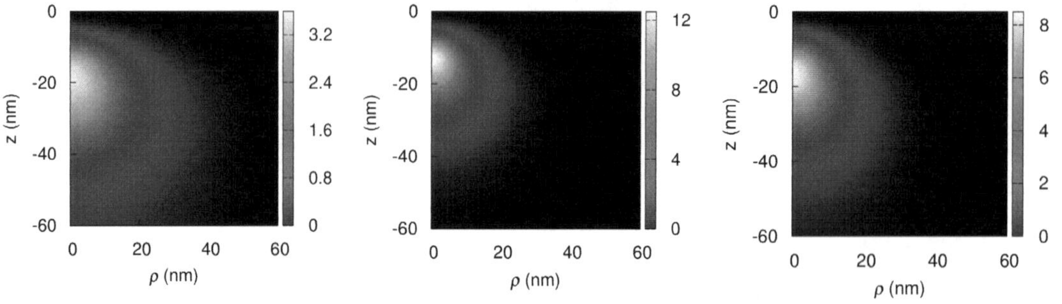

Figure 2. $|F(\rho, z)|^2$ (10^{-5} nm^{-3}) for a Si dopant at $z_d = -0.6$ nm. Left: calculation (i) without image potentials in the absence of a tip. Centre: calculation (ii) including image potentials in the absence of a tip. Right: calculation (iii) including image potentials and TIBB ($V_{tip} = -0.5$ V, $\boldsymbol{\rho}_t = 0$, $z_t = 0.5$ nm).

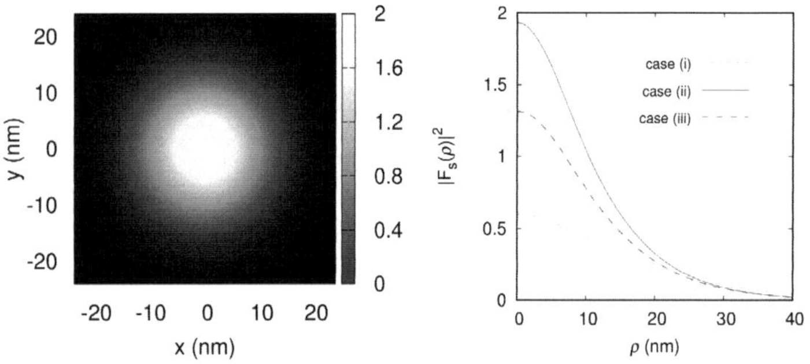

Figure 3. Surface envelope function $|F(\rho, z)|^2$ (10^{-4} nm^{-2}). Left: simulated STM image, $I \propto |F_s|^2$, for case (ii). Right: line scans through $|F_s|^2$ showing the effect of the image potentials and the TIBB on the image. Cases (i),(ii),(iii) are discussed in the text.

In Fig. 3 we can see the effect of the TIBB on the simulated image is significant and this illustrates a challenge when interpreting STM images of weakly bound states: the state is affected by the tip and so the character of the state that is imaged will depend on the position of the tip as it scans across the image. In principle, to simulate the image we should recalculate v_{bb}, $|F(\rho, z)|^2$ and $|F_s|^2$ for each separate tip position.

In summary, we have simulated STM images of the state localised around an Si dopant 0.6 nm below a GaAs surface. We have shown that the tip induced band bending can have a significant effect on the imaged LDOS and find that the effect of the image potentials is to increase the binding energy of the sub-surface state from 1.4 to 3.2 meV. Although this increase in binding energy is significant it is not enough to explain the experimentally observed enhancement.

References

[1] Roy M, Maksym P A, Bruls D, Offermans P and Koenraad P M, Phys. Rev. B **19**, 195304 (2010).
[2] Wijnheijmer A P, Garleff J K, Teichmann K, Wenderoth M, Loth S, Ulbrich R G, Maksym P A, Roy M, and Koenraad P M, Phys. Rev. Lett. **102**, 166101 (2009).
[3] Teichmann K, Wenderoth M, Loth S, Ulbrich R G, Garleff J K, Wijnheijmer A P, and Koenraad P M, Phys. Rev. Lett. **101**, 076103 (2008).
[4] Tersoff J and Hamann D R, Phys. Rev. Lett. **50**, 1998 (1983).
[5] $w_{00}(\boldsymbol{\rho}, z)$ is the tail of the Bloch state above the surface. It can be easily found from a small scale ab-initio calculation of the host semiconductor in the absence of the nanostructure or impurity.
[6] Feenstra R M, J. Vac. Sci. Technol. B **21**, 2080 (2003).
[7] Hao Y L, Djotyan A P, Avetisyan A A, and Peeters F M, Phys. Rev. B **80**, 035329 (2009).
[8] Roy M, and Maksym P A, Phys. Rev. B **68**, 235308 (2003).

Journal of Physics: Conference Series **526** (2014) 012009 doi:10.1088/1742-6596/526/1/012009

LDA DFT simulations of an isolated silicon donor on the (110) surface of GaAs

F J Tilley, Mervyn Roy and P A Maksym

Department of Physics and Astronomy, University of Leicester, University Road, Leicester LE1 7RH, UK

E-mail: fjt3@le.ac.uk

Abstract. The convergence of the band gap state of a single silicon dopant on the (110) surface of GaAs was investigated. By simulating different sized super-cells we were able to show that a 3×4 super-cell provides a well converged calculation for modelling an isolated dopant, with the total energy being converged to 1 part in 1000. The local density of the silicon band gap state was then checked against a number of more intensive calculations and was found to be well converged, with an eigenvalue accurate to within 3 meV.

1. Introduction

Recent STM experiments have shown interesting properties for silicon on the GaAs (110) surface, including positional bistability [1] and enhanced binding energy [2]. We have performed LDA DFT calculations for an isolated silicon donor on the (110) surface of GaAs, and by looking in detail at the localised band gap state of the dopant we can obtain a direct link with the STM results [1, 2] and other theoretical work [3]. This is an important first step in understanding the properties of all group IV dopants on the GaAs (110) surface.

In order to be confident in any theoretical simulation a host of convergence checks must be employed. When simulating isolated donors it is particularly important to check the convergence with respect to the size of the super-cell. Larger cells provide better approximations to isolated dopant systems, but computational costs place upper limits on what can realistically be modelled. A common choice of system size for looking at donors on the (110) surface is a 3×2 lateral super-cell [3] and work has also been done on systems as small as 2×2 [4], but to our knowledge no systematic tests have been reported to show the validity of cells of these sizes.

For this system we are interested in the highest occupied molecular orbital (HOMO). We calculate the total energy and the HOMO energy for the system and converge these with respect to E_{cut}, **k**-point grid and the number of atoms in the system. Through the simulation of more computationally costly systems we show that, in addition to the energies, the LDOS for the HOMO is also well converged.

2. Computational Details

We used the ABINT software package [5] to perform LDA DFT calculations of the (110) surface of GaAs. The psuedopotentials used for the GaAs slab and the silicon dopant were taken from the set developed by the Fritz-Haber Institute which are available through the ABINIT website.

We investigated the dependence on super-cell size using twelve super-cells, all consisting of 7 atomic layers and 12 Å of vacuum but varying in their in-plane dimensions. The super-cells were: 2×2 (56 atom), 2×3 and 3×2 (84 atoms), 2×4 and 4×2 (112 atoms), 3×3 (126 atoms), 4×3 and 3×4 (168 atoms), 5×2 and 2×5 (140 atoms) and 5×3 and 3×5 (210 atoms). For all cells up to 126 atoms the silicon atom replaced a gallium atom in the surface layer of the slab and the entire system was allowed to fully relax, with a tolerance on the relaxation of 1×10^{-4} Hartrees per Bohr. The super-cells larger than 126 atoms were constructed from the final positions of the relaxed 126 atom super-cell with the additional atoms added using the relaxed geometry of the clean GaAs slab. To improve the speed of the relaxation procedure the initial position of the silicon was calculated using a simple geometrical model that treats the atoms as close-packed, hard-walled spheres with a radius equal to the experimentally measured covalent radius. For each super-cell we checked the convergence with respect to E_{cut} and **k**-point grid and across all the different cells we looked at the behaviour of the total energy and the energy of the HOMO.

3. Results and Discussions

The results of the convergence tests can be seen in figure 1. As a result of these checks we developed a standard calculation of the silicon dopant which had a cut-off energy of 30 Ry and a 2×2×1 **k**-point grid. These values of E_{cut} and **k**-grid provided a convergence on the total energy of the system of 1 part in 1000 and a convergence of the HOMO energy of 1 part in 5000 (1 meV) for each super-cell size. This high level of convergence on the HOMO removed any source of uncertainty when it came to looking at the state's energy variation as a result of the changing system size.

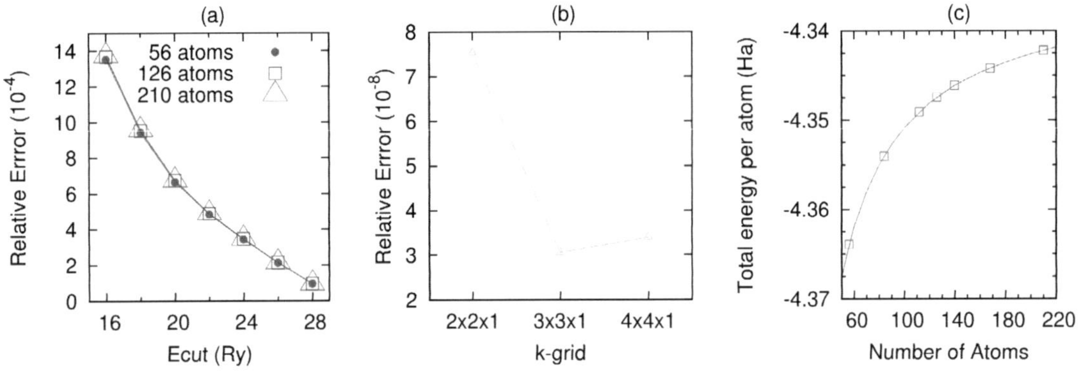

Figure 1. Convergence checks for the total energy of the system. (a) Relative error in E_{total}, as a function of E_{cut}, for three different super-cell sizes as measured from a system with a 30 Ry E_{cut}. (b) Relative error in E_{total}, as a function of **k**-point grid, as measured from a 5×5×1 **k**-grid system. (c) E_{total} per atom as a function of super-cell size.

In figure 2 we show the variation of the HOMO energy with respect to system size. Panels (a) and (b) indicate that the variation in the energy of the dopant state is much more sensitive to changes in the x dimension of the super-cell compared to changes in y. Taking the 210 atom cell to be the converged case we find in (c) that the super-cells with a width of 2 in the x direction produce energies considerably below the converged level regardless of the length in y. This is because the silicon dopant is only 7.82 Å away from its closest periodic image and so a sizeable silicon-silicon interaction is expected. Once the x separation has changed to 3 in (d) we see the energy level settle on the converged value quite rapidly. From these results we chose the 3×4 cell as the size for our standard calculations, giving us a HOMO energy that was converged to within 10 meV, with respect to system size, of its value in the 210 super-cell.

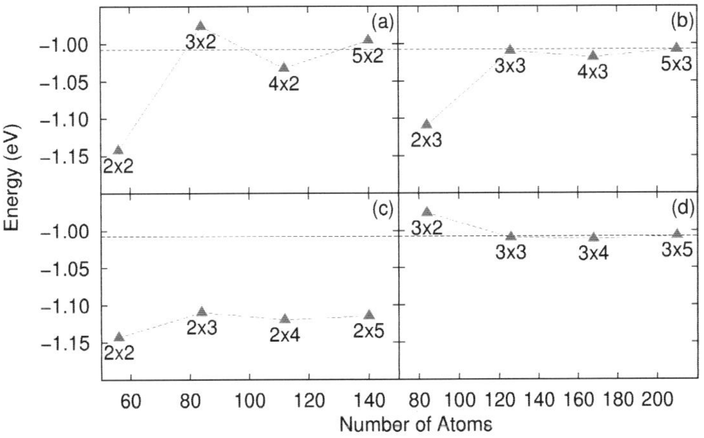

Figure 2. Variation in the energy of the HOMO with lateral super-cell size. (a) Varying the x dimension while keeping y fixed at 2, (b) same but with y fixed at 3, (c) varying y with x fixed at 2 and (d) same as (c) but with x now fixed at 3. In each case the dimensions of the super-cell are given. The horizontal dashed line marks the converged energy level.

The different behaviour of the energy for variations in x and y is primarily caused by the non-equivalence in the lengths of the primitive (110) cell; the y repeating length is $\sqrt{2}$ times longer than x. This causes increases in y to separate the periodic dopants much more rapidly than for changes in x. Another factor is that the (110) slab has zig-zag chains in the plane of the surface in the x direction which may lead to greater defect-defect interactions than are possible in the y direction [6].

Once we had converged the energy of the system we were able to check the robustness of the calculation by simulating the LDOS for a range of different input parameters. Three of the additional simulations involved increasing numerical variables of the calculation while the fourth was run with a different choice of pseudopotential. The different pseudopotentials used were taken from the Hartwigsen-Goedecker-Hutter (HGH) library on the ABINIT website. Figure 3 (a) shows the silicon LDOS for the standard system, a 3×4 cell with the cut-off energy and **k**-point grid described at the start of section 2. We also present the LDOS for the four other systems used to check its robustness. We show the result of the simulation with the different pseudopotentials (b), increasing E_{cut} to 34 Ry (c), the **k**-point grid being increased to a 3×3×1 grid (d), and a system where the number of atoms was increased to 210 (e).

In figure 4 we have taken a line scan of the LDOS through the slab at the location of the

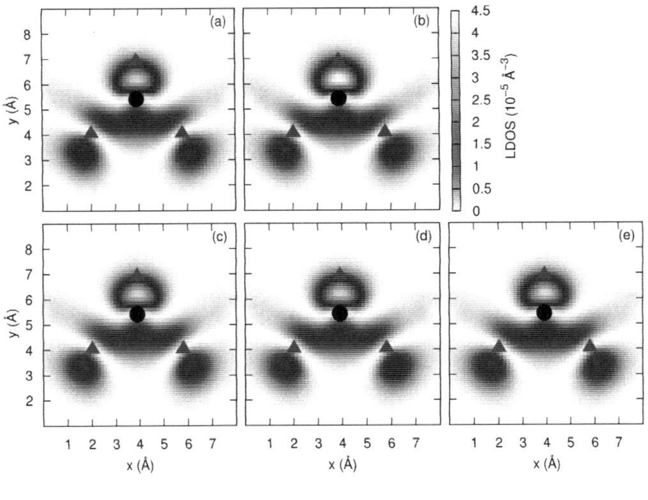

Figure 3. Comparison of LDOS at the silicon location for (a) standard calculation, (b) HGH pseudopotential, (c) E_{cut}=34 Ry, (d) 3×3×1 **k**-point grid and (e) 210 atom super-cell. Unless stated otherwise all other variables match the standard calculation.

Journal of Physics: Conference Series **526** (2014) 012009 doi:10.1088/1742-6596/526/1/012009

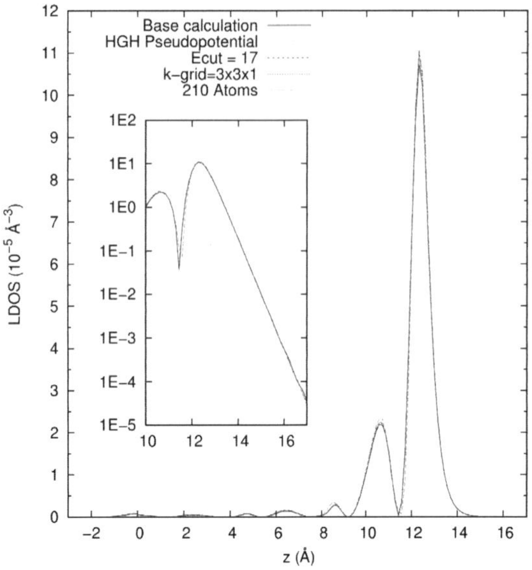

Figure 4. Comparison of line scans of the LDOS taken in z though the position of the silicon dopant for the 5 different simulations. Inset: Log scale plot of the data in the range of 10 to 17 Å. The edges of the relaxed slab are at -0.3 and 12 Å.

silicon. Here we can clearly see the agreement between the five different simulations, indicating that our standard values of E_{cut}, **k**-points, super-cell size and pseudopotential lead to a very well converged simulation, with the peak of the density agreeing to better than 1 part in 30 in all cases. The inset in figure 4 shows the decay of the density into the vacuum on a log scale, indicating the continuation of the agreement into the vacuum tail of the density.

We have also performed the analysis presented here for three other group IV dopants; carbon, germanium and tin. Similar convergence behaviour with respect to super-cell size was seen and the same level of HOMO convergence was obtained with the standard calculation, however for carbon an E_{cut} of 36 Ry was needed.

In conclusion, we have presented our results for a single silicon dopant on the (110) surface of GaAs. For a 168 atom super-cell with an E_{cut} of 30 Ry and a $2{\times}2{\times}1$ **k**-point grid we have shown that our calculation provides a simulation of the silicon HOMO with a well converged LDOS and with an eigenvalue that is converged to within 3 meV.

References

[1] Wijnheijmer, A. P. and Garleff, J. K. and Teichmann, K. and Wenderoth, M. and Loth, S. and Koenraad, P. M. 2011 *Phys. Rev.* B **84** 125310

[2] Wijnheijmer A P, Garleff J K, Teichmann K, Wenderoth M, Loth S, Ulbrich R G, Maksym P A, Roy M and Koenraad P M 2009 *Phys. Rev. Lett.* **102** 166101

[3] Yi, Zhijun and Ma, Yuchen and Rohlfing, Michael 2011 *J. Phys. Chem.* C **115** 23455–62

[4] Wang, Jing and Arias, T. A. and Joannopoulos, J. D. and Turner, G. W. and Alerhand, O. L. 1993 *Phys. Rev.* B **47** 10326–34

[5] The ABINIT code is a common project of the Universit Catholique de Louvain, Corning Incorporated, and other contributors `http://www.abinit.org`.

[6] Lany S and Zunger A 2009 *Modelling Simul. Mater. Sci. Eng.* **17** 084002

Electronic states of elongated PbSe/PbS Core/shell quantum dots

T Walsh[1], J Miloszewski[1], U Aeberhard[2] and S Tomić[1]

[1]University of Salford, Greater Manchester, United Kingdom
[2]IEK-5 Photovoltaik, Forschungszentrum Jülich, Germany

Abstract. The optical characteristics of colloidal quantum dots (QDs) are highly dependent on the physical geometry of the QD (size, shape) as well as composition. These dependencies make such systems attractive for application in novel optical devices, notably for solar cell technology. Empirical electronic structure methods, such as $\boldsymbol{k}\cdot\boldsymbol{p}$ theory, or empirical pseudopotential theories have successfully reproduced experimentally observed transitions in CdSe and PbSe colloidal QDs. Our approach uses the $\boldsymbol{k}\cdot\boldsymbol{p}$ method to predict such properties as the electronic structure and dipole transitions of ellipsoidal PbSe/PbS core/shell structure colloidal QDs, as a function of eccentricity. Due to the anisotropy between the longitudinal (z) and transverse (x and y) directions, we present results from elongation along both the x and z directions.

1. Introduction

The current generation of solar cell devices, based on single gap bulk semiconductor materials, have a maximum theoretical efficiency, known as the Shockley-Queisser limit [1], based on the principle of detailed balance, typically in the range of 30%-40%. It was suggested that nanocrystal quantum dots, also known as "artificial atoms", may be used to overcome this limit [2]. The tunable nature of the QDs (through growing crystals of appropriate size, shape and composition) allows for the selection of band gap energies, as well as associated electronic and optical properties. It is therefore desirable to develop a theoretical framework with which to predict and explore these properties. In the current work, we present our findings on the effect of eccentricity on electronic and optical properties for PbSe/PbS core/shell colloidal QDs, for elongation in both longitudinal (z) and equivalent transverse (x, y) directions, along with the methodology used when performing calculations.

2. Methodology

In the non-interacting electron level of theory, the four–band $\boldsymbol{k}\cdot\boldsymbol{p}$ Hamiltonian, expanded in the $|L_6^-;\alpha\rangle$, $|L_6^+;\beta\rangle$, $|L_6^+;\alpha\rangle$, $|L_6^-;\beta\rangle$ basis states in the vicinity of the L characteristic point in the first Brillouin zone of the rock-salt crystal structure, is given as [3]:

$$H = \begin{vmatrix} |L_6^-;\alpha\rangle & |L_6^-;\beta\rangle & |L_6^+;\alpha\rangle & |L_6^+;\beta\rangle \\ E_g + \frac{\hbar^2 k_\parallel^2}{2m_\parallel^-} + \frac{\hbar^2 k_z^2}{2m_z} & 0 & \frac{\hbar}{m}P_z k_z & \frac{\hbar}{m}P_\parallel(k_x - ik_y) \\ 0 & E_g + \frac{\hbar^2 k_\parallel^2}{2m_\parallel^-} + \frac{\hbar^2 k_z^2}{2m_z} & \frac{\hbar}{m}P_\parallel(k_x + ik_y) & -\frac{\hbar}{m}P_z k_z \\ \frac{\hbar}{m}P_z k_z & \frac{\hbar}{m}P_\parallel(k_x - ik_y) & E_g + \frac{\hbar^2 k_\parallel^2}{2m_\parallel^-} + \frac{\hbar^2 k_z^2}{2m_z} & 0 \\ \frac{\hbar}{m}P_\parallel(k_x - ik_y) & -\frac{\hbar}{m}P_z k_z & 0 & E_g + \frac{\hbar^2 k_\parallel^2}{2m_\parallel^-} + \frac{\hbar^2 k_z^2}{2m_z} \end{vmatrix} \quad (1)$$

The anisotropy between the longitudinal (z), $\langle 111\rangle$, and transverse ($\parallel$) directions of the rock-salt crystal structure (x and y are taken respectively along the $\langle\bar{1}\bar{1}2\rangle$ and $\langle 1\bar{1}0\rangle$ crystallographic directions) is accounted for within the difference in the effective masses of the electrons, $m_{\pm,z}^*$ and $m_{\pm,\parallel}^*$, and the bulk dipole matrix elements, $P_{0,z}$ and $P_{0,\parallel}$ [3]. The polarization-dependent optical matrix elements, $|\hat{\boldsymbol{e}}\cdot\boldsymbol{p}_{if}|^2$, of the QD required for the description of radiative transitions are obtained using the *Hellman-Feynman theorem*, where $\hat{\boldsymbol{e}}$ is the unit light polarization vector, and $\boldsymbol{p}_{if}(\boldsymbol{k}) = (m_0/\hbar)\langle i|\partial H_k/\partial k|f\rangle$ is the electron-hole momentum matrix element of the structure, with $|i\rangle$ and $|f\rangle$ being the envelope functions of the initial and final states of the radiative process respectively. In order to capture the variation in material parameters between the core, shell, and surrounding colloid the radial variation of $P_{0,\|,z}$ and $m_{0,\|,z}$ are included, as well as the variation in the band edge energy of the conduction band minimum and valence band maximum. To assess the absorption properties of various QD structures we use the expression for the absorption cross section:

$$\sigma(\hbar\omega) = \frac{\pi}{nc\epsilon_0\omega}\left(\frac{e}{m_0}\right)^2 \sum_{if} \left|\hat{\boldsymbol{e}} \cdot \boldsymbol{p}_{if}\right|^2 \delta\left(E_f - E_i - \hbar\omega\right),$$

given in units of cm^2, where ϵ_0 is the permittivity of the free space, e is the electron charge, and ω is the incident light frequency. The delta function, δ, is replaced with a Gaussian function $\exp\left[-\left(\frac{x}{\sqrt{2}\Delta}\right)^2\right]/(\sqrt{2\pi}\Delta)$, defined by the phenomenological line broadening, Δ, set to x,y,z = 25meV in all structures considered. All results presented here are obtained with kppw code [4].

3. Results

The **k·p** Hamiltonian in the x- and y-directions are equivalent due to rock-salt crystal symmetry, however he z-direction produces different properties due to crystal anisotropy. The shape of the QD can break Hamiltonian symmetry. Starting from the spherical QD shape, we calculate the optical dipole matrix elements, electronic structure, wave functions, and absorption cross-sections of PbSe/PbS core-shell QDs of increasing eccentricity, via independent elongation either along the x- or z-directions. If the QD is elongated along the x-direction, keeping y- and z- radii the same, the QD shape breaks the in-plane isotropy of the **k·p** Hamiltonian. However if the QD is elongated along the z-direction, keeping x- and y- radii the same, such a shape only exaggerates the already existing anisotropy of the **k·p** Hamiltonian.

3.1. Energy levels

The elongation along a single axis increases the volume of the QD, resulting in a red shift in the energy spectra. The magnitude of this energy shift is different depending on the direction of elongation (Figure 1). For spherical PbSe/PbS QDs of core/shell diameter 15Å/25Å, the characteristic $e_0 - h_0$ transition is found to occur at an energy of 1.69eV. Elongation of the QD to 30Å/50Å in only the x (or equivalently y) direction results in a lowering of the band gap energy to 1.12eV, while the same elongation in only the z-direction reduces the band gap energy to 1.01eV. For a spherical QD of core/shell diameter 30Å/50Å, E_g = 0.80eV. This red shift of the $e_0 - h_0$ transition is a consequence of a simple increase of the confinement volume. The reduction in transition energies allows for absorption of photons of lesser energy, with the $e_0 - h_0$ optical gap energy being indirectly proportional to the dot size, allowing tuning of the absorption edge energy through choosing QDs of appropriate size. The lower $e_0 - h_0$ transitions allow for greater use of the solar spectrum, and for a potential increased efficiency of the solar cell device.

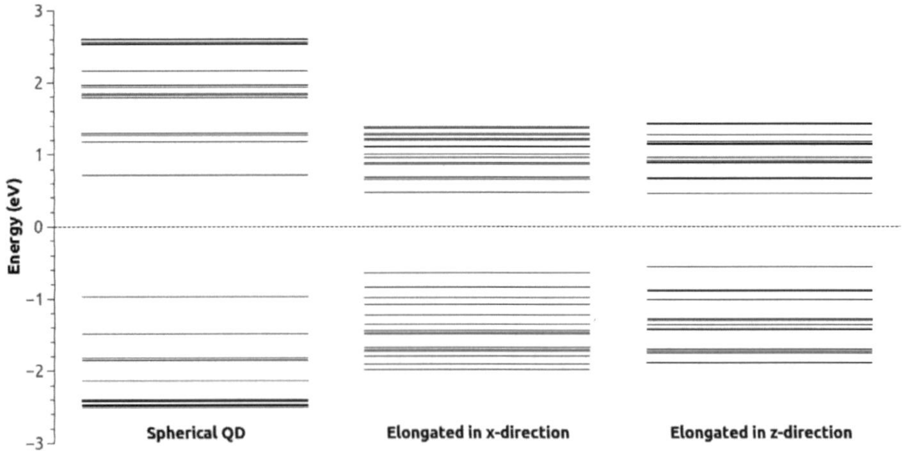

Figure 1: Energy levels for PbSe/PbS core/shell QD. Left to right, spherical, elongated in x-direction, elongated in z-direction.

3.2. Dipole elements

For the spherical core/shell structure QD, the equivalent x- and y-polarisations have identical dipole moments, occurring at the same transition energy (Figure 2), due to the isotropic crystal structure in these directions. The z-component has a lesser dipole moment. Furthermore, elongation in the x-direction causes a break in overall symmetry, resulting in an anisotropic system Hamiltonian, and different dipole moments for the x- and y-polarisations. It is also apparent there are many more optically allowed/non-vanishing transitions in the elongated samples. They are a consequence of overall symmetry breaking induced by the elongation of the QDs in certain directions (Figure 2 (b), (c)).

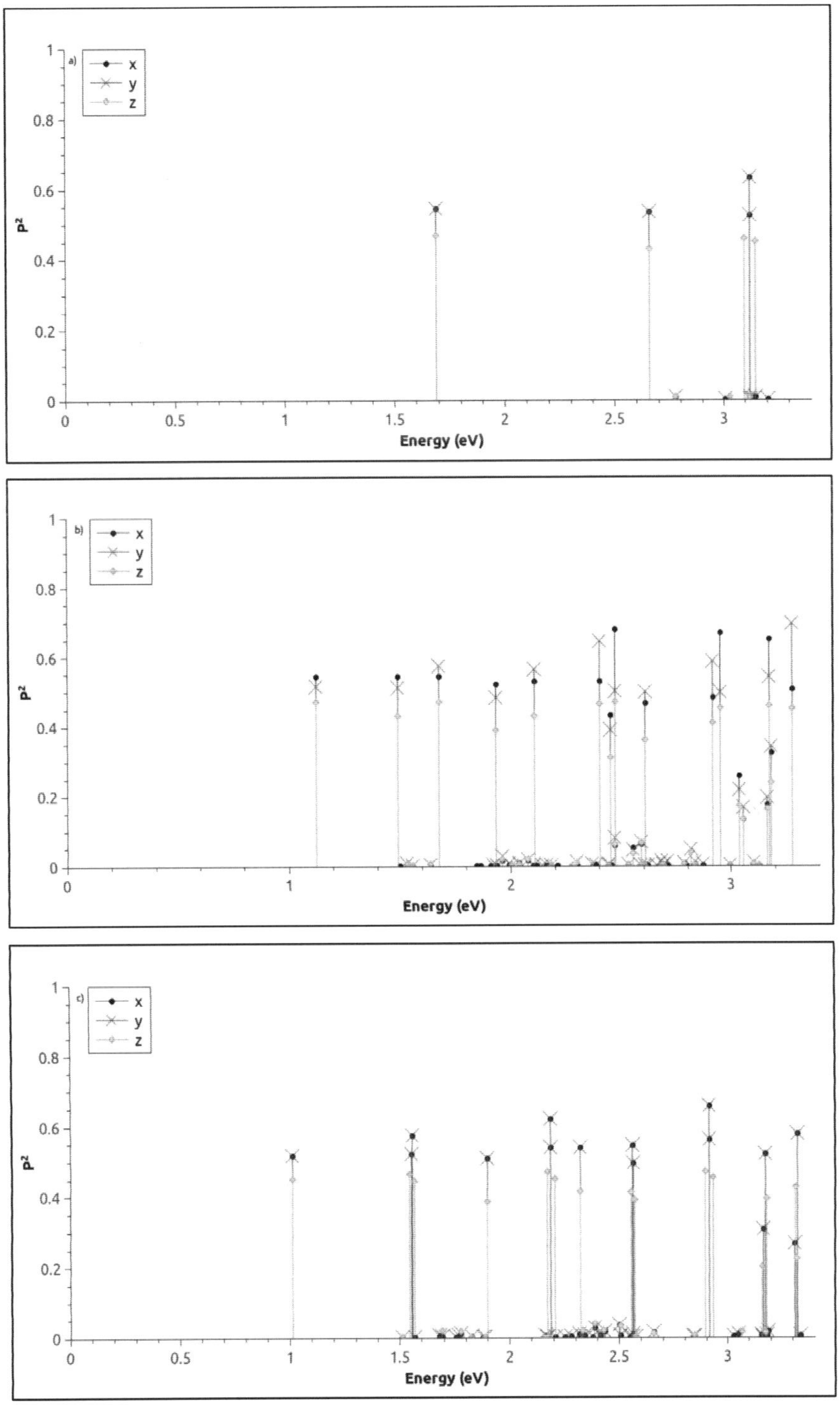

Figure 2: Dipole transitions for (a) spherical core/shell QD of diameter 15Å/25Å, (b) QD elongated in x-direction only to core/shell diameter of 30Å/50Å, and (c) QD elongated in z-direction only to core/shell diameter of 30Å/50Å.

3.3. Absorption spectra

In an ensemble of QDs, different core/shell radii are present, thus QD properties should be considered for QDs of a significant size distribution [5]. The absorption cross-section of each QD in one such distribution is shown in Figure 3. We consider nine particular QDs in the size range of core diameter 15Å-30Å and shell diameter 25Å-50Å. Figure 3 (a) represents the absorption cross-sections of QDs elongated in the x-direction with increment 1.875Å for the core and 3.125Å for the shell, while Figure 3 (b) represents cross-sections of QDs elongated in the z-direction by the same increment.

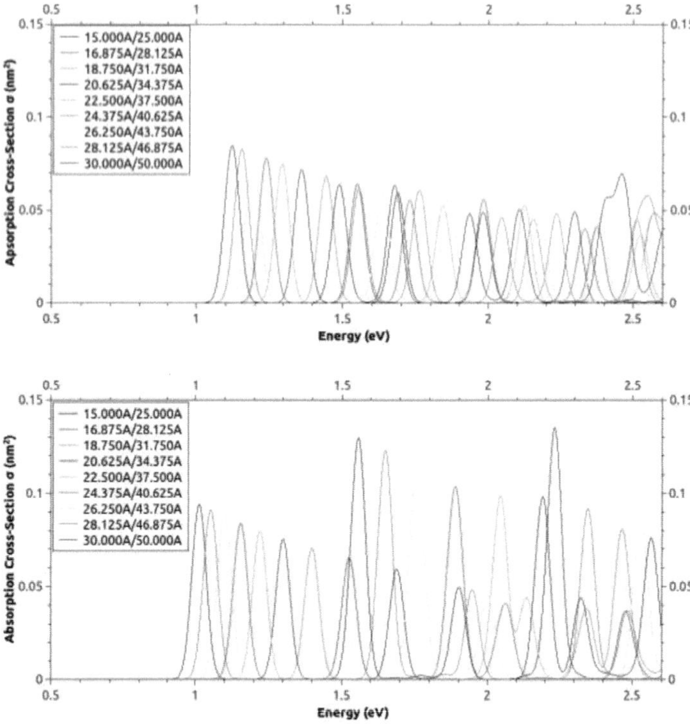

Figure 3: Absorption cross-sections in z-polarisation for QD elongation in x (top) and in z (bottom) for sample of nine QDs with core diameters 15Å-30Å and shell diameters 25Å-50Å.

An increase in dot volume results in a shift of the absorption cross-section peaks to lower energies, with a larger shift for elongation in the z-direction than for elongation in the x- or y-directions (0.68eV in the z-direction as compared to 0.57eV in the x-direction for the largest QDs). The magnitude of the absorption peaks also increases with dot volume, allowing for greater probability of absorption of lower energy photons.

4. Conclusions

The effects of eccentricity on the optical properties of ellipsoidal colloidal QDs are highly pronounced, with the direction of elongation playing a key role. Elongation in the x- or equivalent y-direction ($\langle\bar{1}\bar{1}2\rangle$ or $\langle 1\bar{1}0\rangle$ crystallographic direction) from 15Å/25Å to 30Å/50Å results in a significant reduction in optical gap energy of 0.57eV. Elongation of only the x- or y-direction introduces anisotropy, with the break in symmetry causing the x and y dipole moments to become non-identical (as compared with the in-plane isotropic spherical case). Elongation along the *z*-direction ($\langle 111\rangle$ direction) results in a larger reduction of the optical band gap energy of 0.82eV, along with an increase in the absorption cross-sections.

5. Acknowledgements

The authors wish to gratefully acknowledge the EPSRC Doctoral training account (DTA) for their funding in this project.

References

[1] W. Shockley and H. Queisser, J. Appl. Phys. **32**, 510 (1961)
[2] J. An et al., Nano Lett. **6**, 2728 (2006)
[3] I. Kang and F. W. Wise, J. Opt. Soc. Am. B **14**, 1632 (1997).
[4] S. Tomić. A. G. Sunderland and I. J. Bush, J. Mater. Chem. **16**, 1963 (2006)
[5] U. Aeberhard, R. Vaxenburg, E. Lifshitz and S. Tomić, Phys. Chem. Chem. Phys. **14**, 16223 (2012)

AUTHOR INDEX